Freshwater Aquatic Biomes

GREENWOOD GUIDES TO BIOMES OF THE WORLD

Introduction to Biomes
Susan L. Woodward

Tropical Forest Biomes
Barbara A. Holzman

Temperate Forest Biomes
Bernd H. Kuennecke

Grassland Biomes
Susan L. Woodward

Desert Biomes
Joyce A. Quinn

Arctic and Alpine Biomes
Joyce A. Quinn

Freshwater Aquatic Biomes
Richard A. Roth

Marine Biomes
Susan L. Woodward

Freshwater Aquatic
BIOMES

Richard A. Roth

Greenwood Guides to Biomes of the World

Susan L. Woodward, General Editor

GREENWOOD PRESS

Westport, Connecticut • London

Library of Congress Cataloging-in-Publication Data

Roth, Richard A., 1950–
    Freshwater aquatic biomes / Richard A. Roth.
        p. cm. — (Greenwood guides to biomes of the world)
    Includes bibliographical references and index.
        ISBN 978-0-313-33840-3 (set : alk. paper) — ISBN 978-0-313-34000-0
(vol. : alk. paper)
    1. Freshwater ecology. I. Title.
    QH541.5.F7R68 2009
    577.6—dc22   2008027511

British Library Cataloguing in Publication Data is available.

Library of Congress Catalog Card Number: 2008027511
ISBN: 978-0-313-34000-0 (vol.)
        978-0-313-33840-3 (set)

First published in 2009

Greenwood Press, 88 Post Road West, Westport, CT 06881
An imprint of Greenwood Publishing Group, Inc.
www.greenwood.com

Printed in the United States of America

The paper used in this book complies with the
Permanent Paper Standard issued by the National
Information Standards Organization (Z39.48–1984).

10 9 8 7 6 5 4 3 2 1

Contents

Preface

This volume describes the freshwater aquatic biome, which consists of lakes, rivers, and wetlands. These life zones are distinguished from terrestrial biomes, such as deserts and tropical forests, and from the marine biome. They thus occupy a unique place in the biosphere. That said, as is the case with other biomes, our conceptual categories are much neater than living nature, which is much more likely to have fluctuating gradients rather than sharp dividing lines. Thus, for example, freshwater and saltwater tidal marshes exist along a continuum of salinity; riparian wetlands may be part of the river at times. Nonetheless, our use of concepts and categories helps us to make sense of the world, and in this volume, many concepts applicable to freshwater systems are introduced.

Just as this series follows the conventional biogeographic division of Earth's living systems into the major biomes, I have followed standard practice in categorizing the freshwater aquatic biome into the three major categories of rivers, lakes, and wetlands. One type of life environment that does not fit easily into any of these three freshwater environments is salt lakes. They are not freshwater environments; nonetheless they are included in this volume, because, one might say, a salt lake is more like a lake than like the ocean.

In each of the major freshwater aquatic environments, three examples are presented in some depth. In each case, I describe a low-, a mid-, and a high-latitude system. While this approach is a little different from that followed in the volumes on terrestrial biomes, it offers a broader range of specific manifestations of freshwater aquatic environments. For example, lakes at very different latitudes are likely to encompass a greater range of physical conditions than lakes at the same latitude on different continents or in different biogeographic realms.

In the chapters on rivers, lakes, and wetlands, I spend considerable time explaining the range of physical conditions within which life has evolved in these environments. I also describe how lifeforms have adapted to the conditions. For example, wetland environments are characterized by low oxygen conditions, particularly in the substrate. What adaptations make it possible for plants to survive in such conditions?

Throughout, with an eye toward what I suppose to be the needs and capabilities of the readers of this volume, I have tried to find the right balance between general concepts and specific manifestations. I have attempted to supply enough technical detail to understand a particular environment without unnecessarily burdening the reader.

I thank the series editor, Dr. Susan Woodward, for her assistance, collegiality, good humor, encouragement, and many helpful suggestions.

How to Use This Book

The book is arranged with a general introduction to the freshwater aquatic biome and a chapter on each of the three generally recognized forms of that biome: rivers, wetlands, and lakes. Salt lakes, although not freshwater, are also included in the chapter on lakes, as are manmade lakes (reservoirs). The biome chapters begin with a general overview, proceed to describe the distinctive physical and biological characteristics of each form, and then focus on three examples of each in some detail. Each chapter and each example can more or less stand on its own, but the reader will find it instructive to investigate the introductory chapter and the introductory sections in the later chapters. More in-depth coverage of topics perhaps not so thoroughly developed in the examples usually appears in the introductions.

The use of Latin or scientific names for species has been kept to a minimum in the text. However, the scientific name of each plant or animal for which a common name is given in a chapter appears in an appendix to that chapter. A glossary at the end of the book gives definitions of selected terms used throughout the volume. The bibliography lists the works consulted by the author and is arranged by biome and the regional expressions of that biome.

All biomes overlap to some degree with others, so you may wish to refer to other books among Greenwood Guides to the Biomes of the World. The volume entitled *Introduction to Biomes* presents simplified descriptions of all the major biomes. It also discusses the major concepts that inform scientists in their study and understanding of biomes and describes and explains, at a global scale, the environmental factors and processes that serve to differentiate the world's biomes.

The Use of Scientific Names

Good reasons exist for knowing the scientific or Latin names of organisms, even if at first they seem strange and cumbersome. Scientific names are agreed on by international committees and, with few exceptions, are used throughout the world. So everyone knows exactly which species or group of species everyone else is talking about. This is not true for common names, which vary from place to place and language to language. Another problem with common names is that in many instances European colonists saw resemblances between new species they encountered in the Americas or elsewhere and those familiar to them at home. So they gave the foreign plant or animal the same name as the Old World species. The common American Robin is a "robin" because it has a red breast like the English or European Robin and not because the two are closely related. In fact, if one checks the scientific names, one finds that the American Robin is *Turdus migratorius* and the English Robin is *Erithacus rubecula*. And they have not merely been put into different genera (*Turdus* versus *Erithacus*) by taxonomists, but into different families. The American Robin is a thrush (family Turdidae) and the English Robin is an Old World flycatcher (family Muscicapidae). Sometimes that matters. Comparing the two birds is really comparing apples to oranges. They are different creatures, a fact masked by their common names.

Scientific names can be secret treasures when it comes to unraveling the puzzles of species distributions. The more different two species are in their taxonomic relationships, the farther apart in time they are from a common ancestor. So two species placed in the same genus are somewhat like two brothers having the same father—they are closely related and of the same generation. Two genera in the same family might be thought of as two cousins—they have the same grandfather, but different fathers. Their common ancestral roots are separated farther by time. The important thing in the study of biomes is that distance measured by time often means distance measured by separation in space as well. It is widely held that new species come about when a population becomes isolated in one way or another from the rest of its kind and adapts to a different environment. The scientific classification into genera, families, orders, and so forth reflects how long ago a population went its separate way in an evolutionary sense and usually points to some past environmental changes that created barriers to the exchange of genes among all members of a species. It hints at the movements of species and both ancient and recent connections or barriers. So if you find two species in the same genus or two genera in the same family that occur on different continents today, this tells you that their "fathers" or "grandfathers" not so long ago lived in close contact, either because the continents were connected by suitable habitat or because some members of the ancestral group were able to overcome a barrier and settle in a new location. The greater the degree of taxonomic separation (for example, different families existing in different geographic areas), the longer the time back to a common ancestor and the longer ago the physical separation of the species. Evolutionary history and Earth history are hidden in a name. Thus, taxonomic classification can be important.

Most readers, of course, won't want or need to consider the deep past. So, as much as possible, Latin names for species do not appear in the text. Only when a common English language name is not available, as often is true for plants and animals from other parts of the world, is the scientific name provided. The names of families and, sometimes, orders appear because they are such strong indicators of long isolation and separate evolution. Scientific names do appear in chapter appendixes. Anyone looking for more information on a particular type of organism is cautioned to use the Latin name in your literature or Internet search to ensure that you are dealing with the correct plant or animal. Anyone comparing the plants and animals of two different biomes or of two different regional expressions of the same biome should likewise consult the list of scientific names to be sure a "robin" in one place is the same as a "robin" in another.

1

Introduction to Freshwater Aquatic Biomes

Freshwater Aquatic Environments as a Biome

The freshwater aquatic biome must be included in any complete treatment of the Earth's living environments. Yet, because the biome concept was developed as a way to understand and categorize *terrestrial* environments, freshwater aquatic and marine environments do not easily fit the same template set forth for land-based systems. Climate is certainly a factor in aquatic ecosystems, but locally specific conditions such as water chemistry, hydrologic regime, type and frequency of disturbance, and geologic history also determine the nature of the biota and its interrelationships with living and nonliving aspects of the habitat, including surrounding terrestrial habitat.

In this volume, the three major types of freshwater habitats—lakes, rivers, and wetlands—are each given a separate chapter. In this introductory chapter, some physical and biological aspects of the freshwater aquatic biome common to all three are presented. For specific details on physical and biological aspects of wetlands, lakes, and rivers, see the following chapters.

The Interconnectedness of Freshwater Aquatic Systems

In this book, we consider wetlands, rivers, and lakes individually as if they were separate and distinct entities. In the real world, however, the distinctions are blurred. As with so much of human knowledge of Earth systems, descriptions necessarily simplify. The real world is messy and complex. Sharp lines, such as the conceptual distinction between a river and a wetland, or the boundary on a map

1

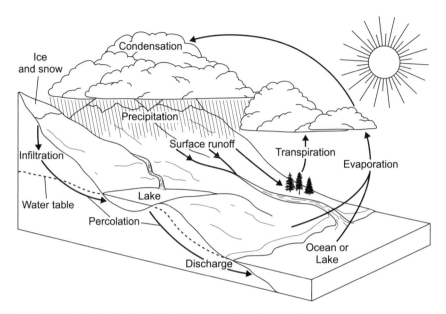

Figure 1.1 The hydrologic cycle. *(Illustration by Jeff Dixon.)*

between a lake and its upland surroundings, are seldom seen in nature. We simplify because it helps us to understand, and it is useful as long as we do not forget that our models are simplifications.

Clearly all surface water (and much groundwater as well) is part of the Earth's hydrologic cycle (see Figure 1.1). Beginning arbitrarily with the point at which water evaporates, the cycle works as follows: Water evaporates from the oceans, meaning that it goes from a liquid state to a gaseous state. Some of this evaporated water is transported by wind currents in the form of clouds over land, where it may form into droplets (liquid) or crystals (solid) and precipitate onto the land. It may come down in the familiar forms of rain, snow, and sleet, as well as heavy fog, rime ice, or frost.

Precipitation lands on the ground surface, trees, rooftops, and roads. Once it lands (or, in the case of snow and ice, once it melts) it may run off, infiltrate into the ground, or evaporate and return to the atmosphere. The subject of this volume, the freshwater aquatic biome, is largely concerned with water that has either run off or infiltrated. Runoff water invariably moves downhill under the force of gravity, and either collects into a stream system that feeds a river, or collects in surface depressions as ponds, lakes, or wetlands.

Infiltrated water percolates through the soil and subsoil but, like runoff, also moves downhill. Eventually it feeds into a stream or river system; a lake, pond, or wetland; or the sea. Depending on the distance involved, the nature of the subsurface environment (such as gravel deposit, clay layer, sand deposit, or rock), and surface topography (how steep), the journey may take minutes to years or even centuries.

All bodies of water, whether oceans, lakes, ponds, rivers, or wetlands, are connected through their mutual participation in the hydrologic cycle. Whether or not this connection is significant for the biota depends on the particular hydrologic process and on geography.

For example, floodplain wetlands are considered to be wetlands, but they are also properly considered to be part of the river system, even though there may be no direct connection after river levels fall and the river retreats from the floodplain-wetland system. During the flood period when the river and the floodplain become one, river organisms occupy the floodplain and its wetlands. This occupation may play an important part in a particular organism's reproductive cycle. At the same time, the river in flood adds sediment, organic material, and nutrients to the floodplain-wetlands ecosystem. The two systems, which may be considered different biomes, are tightly coupled (see Figure 1.2).

Lakes of any size are fed by surface flows, usually in the form of river systems. Many, like the North American Great Lakes, also feed rivers with their outflow. The hydrologic connection also provides an avenue for nutrient exchanges, sediment movement, and dispersal of living organisms. Lake fishes may spawn in a lake's tributaries. For example, several sucker species in Upper Klamath Lake in the northwestern United States spawn in the Williamson River and the Sprague River. Numerous fish species occupy the remaining wetlands that fringe Upper Klamath Lake during their early life stages, and this is also typical. The biota, over many millennia, have evolved and coevolved to take advantage of the myriad of

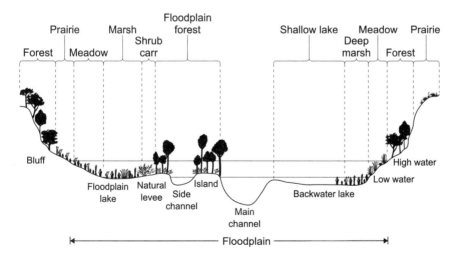

Figure 1.2 A cross-section of a typical river in the North American Midwest. Low river stages prevail most of the time; floods occur less frequently. *(Illustration by Jeff Dixon. Adapted from Theiling et al. 2000.)*

opportunities—for feeding, dispersal, reproduction, and habitation—presented by the interconnected variety of freshwater aquatic environments. In the process, they have collectively developed into the biomes described in this volume.

Unique Conditions to which Organisms Adapt

Life in a water, or aquatic, environment is different in many obvious but important respects than life in a terrestrial environment. In evolutionary terms, a water environment was the original environment: life began in the seas. Terrestrial environments were the foreign territory to which lifeforms had to adapt.

Physical Properties of Water as a Living Environment

As aquatic lifeforms became more complex, their development included many adaptations to the unique conditions of the aquatic environment. The following sections explore some of those conditions and the ways in which aquatic organisms have adapted to them.

Density. One of the physical properties of water is density. Density is defined as mass per unit volume. In the International System of measurement (the SI, also known as the metric system), the density of pure water is 1 gram per cubic centimeter (g/cc). Ocean water, with its high concentration of dissolved solids, has a density of 1.03 g/cc; freshwater is much closer to the density of pure water. Water density is about 800 times that of air at sea level.

The density of water is temperature-dependent. It reaches a maximum at 39.2° F (4° C), just a few degrees above the freezing point. As is explained in more detail in the chapter on lakes, this density difference of water at different temperatures can cause stratification—separation into layers that do not mix—in lakes under some conditions.

Besides the phenomenon of stratification, the density of water makes a difference to organisms living in it in several other ways. Because of its density, water supports the bodies of organisms. Large aquatic animals, whose density is (like ours) not much different from that of water, do not need the heavy musculature and skeletal mass to support them that terrestrial animals do. They have tended to evolve to use their skeletal and muscular masses to support their ability to move in water (which is much more difficult than to move in air). Large aquatic plants do not need rigid trunks and branches capable of supporting heavy loads, as do trees.

Most, but not all, aquatic organisms are slightly denser than water. If the density of an organism is a little less than that of its watery environment, it floats to the surface; if it is a little more, it sinks to the bottom. Therefore aquatic organisms have developed various adaptations to control their position in the water column.

Floating macrophytes (multicellular plants) sometimes have air bladders among their roots to ensure that they stay at the surface. Macrophytes may also be rooted if they are in relatively shallow water. Phytoplankton—one-celled photo-synthesizing plants—tend to be slightly denser than water and therefore tend to sink over time, except in the turbulent waters of streams and rivers. But they need light to survive. They need to stay near the surface, at least during the day, and thus have evolved several different adaptations to solve this problem.

Some phytoplankton can change their density, rather like a hot-air balloonist raising and lowering a balloon. Some have gas vesicles or vacuoles (gas-filled blad-der-like compartments within their cells) that expand or contract, often in conjunc-tion with the rate of photosynthesis. At the surface during the day, photosynthesis takes place, and as it does, the phytoplankter takes on more mass—like taking on ballast. Pressure within the cell walls increases, causing any gas vesicles to become compressed; relatively dense hydrocarbon molecules are formed, further increasing density. As overall density is increased, the phytoplankter sinks. As it sinks, the rate of photosynthesis drops, and cell respiration takes place. Intracellular pressure is relieved, gas vesicles can expand, and, through the process of respiration, some mass is expelled from the cell (like shedding ballast). Thus, the organism loses den-sity and begins to rise again. Many phytoplankton go through such a cycle on a daily basis in concert with the daily cycle of light and dark.

Another way to control density that has been observed among algae is to secrete a mucus coating that absorbs and holds water; this increases cell size but decreases average density.

Other adaptations help phytoplankton (and zooplankton as well, which need to stay near the surface—that's where their food is!) to maintain optimal positions in the water column. Spherical objects sink relatively rapidly (a sphere has the smallest possible surface area for a given mass), but shapes that depart from the spherical sink less rapidly. An acorn falls faster than a leaf. Phytoplankton can approximate more of a leaf shape by forming colonies. Some algae also sport spines or other feather-like appendages; these are thought to be for the purpose of increasing surface area to resist sinking. Finally, some algae (and many zooplankters) have the ability to propel themselves and can thus adjust their position in the water column. Such algae have an appendage called a flagellum, a whip-like structure that can either be whipped around or rotated in some cases like a propeller, to propel the organism.

The other inhabitants of the water column—fishes mostly—are able to propel themselves effectively and therefore have considerable control over their depth. The largest group of fishes—the ray-finned fish or Actinopterygii—have an internal organ called a gas bladder or swim bladder that they can expand or contract to con-trol their average density and therefore their buoyancy. They do so by filling the bladder with internally produced gas or with air "gulped" at the surface, and then emptying it. Other fishes make themselves more buoyant by storing fats, which are less dense than water.

Viscosity. The problem of movement in water faced by a microscopic creature is related to another property of water, viscosity. Viscosity is defined as the internal resistance to flow. Another way to think of it is as the resistance of water to the movement of an object through it (the physical forces are the same whether the object is moving and the water is stationary or the object is stationary and the water is moving). Viscosity is due to attractive forces between molecules of a fluid; different types of fluids have different types of intermolecular forces. Water has relatively strong intermolecular attraction: water molecules like to stay close. At the same time, water molecules are small, which tends to give a fluid lower viscosity. Most of us experience viscosity as "thickness" of a liquid. We know that molasses and motor oil are both "thicker" than water. What this means is that they do not flow as easily.

The viscosity of water depends to some extent on temperature; it reaches a maximum viscosity just above freezing, and the warmer it is, the lower the viscosity. Attentive canoeists who paddle year-round may notice that their canoes are more difficult to move through the water in winter than in summer.

One way of quantifying the relationship with water faced by a moving object (like a canoe or fish) is the Reynolds number. The Reynolds number precisely describes the balance between viscous forces that resist movement in a fluid, and the force of inertia, by which an object in motion tends to stay in motion. Viscous forces slow an object moving through liquid, while inertial forces keep it moving. At Reynolds numbers greater than 1, inertial forces prevail; less than 1, and viscous forces prevail. An extreme example of a high Reynolds number (astronomically high, actually) might be that of an oil tanker, which will keep moving through water for a long time even after the propeller is turned off. The force of inertia keeps it moving, and the viscous forces of the water, which slow it down, are trivial in comparison. Paddling a canoe, the Reynolds number is lower; the canoeist needs to keep applying force by paddling to maintain velocity.

Reynolds numbers depend in part on the size of the object in question. The smaller the object attempting to move through the water, the lower the Reynolds number. As pointed out by physicist E. M. Purcell, water-dwelling microorganisms, such as the bacteria *E. coli*, experience "life at low Reynolds numbers"—perhaps 10^{-4}. This means, among other things, that if such an organism tried to swim as humans do, using reciprocal motion (that is, moving arms and legs back and forth), it would be like swimming in thick molasses—you wouldn't go anywhere. Microorganisms that can propel themselves through water have evolved a number of interesting means for doing so. Some, like certain species of *E. coli* bacteria, have corkscrew appendages that rotate. Flagella and cilia (short hair-like structures that assist in movement) are other ways of overcoming the difficulties of moving through water experienced by the very small.

Temperature. Water has a high specific heat, defined as the amount of energy it takes to heat 1 gram of a substance by $1°$ C. Certainly compared with air, water resists temperature change. A large body of water, like a large lake or the ocean, makes its surrounding climate less prone to extreme temperatures. Nonetheless, bodies of water do change temperature seasonally in the mid-latitudes, and the organisms that live in them must be able tolerate a range of temperatures.

Many aspects of the aquatic environment change with temperature. Light in the mid-latitudes changes with temperature, as both are dependent on solar radiation. In other words, in the summer, when sunlight is more intense, the temperature is also higher. Density, as described above, changes with temperature, as does viscosity. The ability of substances to dissolve in water is temperature-dependent. The colder the water the more it can hold dissolved gases such as oxygen and carbon dioxide, which are important biologically.

Biological processes such as metabolism, respiration, and photosynthesis tend to have an optimum temperature range, below which they decline, and above which they decline as well. For algae, from about 41° F (5° C) to an optimum around 68° F (20° C), growth rate doubles or more than doubles. Different organisms do well (grow and reproduce) at different temperatures. Cold-water fishes like trout tend to be happiest around 61°–64° F (16°–18° C); warm-water fishes like green sunfish prefer a temperature range about 18° F (10° C) higher.

Light. Photosynthesis is driven by the energy of sunlight; indeed it is a process of capturing and storing that energy. But water is not completely transparent, so with increasing depth, light diminishes as it is absorbed and scattered by water molecules. Light intensity is reduced by a constant percentage per unit depth, which means the decrease with depth is exponential. Freshwater has dissolved and suspended material in it (including plankton) that increases the absorption and scatter and reduces the distance that light can penetrate. Even in exceptionally clear water, below about 330 ft (100 m) sunlight is almost completely gone.

The zone in which there is enough light for photosynthesis to occur is called the euphotic zone (see Figure 1.3). It is defined as that part of the water column from the surface down to the depth at which only 1 percent of the light striking the water surface remains. At this light level, photosynthesis is approximately equal to respiration. The euphotic zone is the only zone in which phytoplankton, or indeed any plant, can live.

Water clarity—its ability to conduct light—is measured in natural surface water bodies with a simple device called a Secchi disk. The disk, which has a highly visible black-and-white pattern, is lowered into the water. The line to which it is attached is marked so the depth at which the disk disappears (the "Secchi depth") can be noted. A rule of thumb is that the euphotic zone extends downward two to three times the Secchi depth.

Chemical Properties of Natural Waters

Water is known as the universal solvent, for its ability to dissolve almost any substance. This means that practically every chemical substance can be found dissolved in water. Some of these are of great importance to freshwater biota.

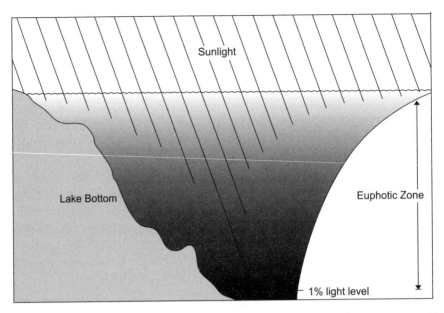

Figure 1.3 The euphotic zone is defined as the part of the water column penetrated by light, extending from the surface down to the depth at which only 1 percent of the original light intensity is present. The diminution of the light is exponential, hence the J-shaped curve. The actual depth at which a particular percentage of light remains will depend on conditions specific to a body of water, such as the amount of dissolved organic material, the density of plankton, and the amount of suspended solids. *(Illustration by Jeff Dixon. Adapted from Burgis and Morris 1987.)*

Salts are a name given to dissolved solids that, upon dissolving, separate into positively and negatively charged particles called, respectively, cations and anions. For example, what most people think of as salt, sodium chloride, separates into a positively charged sodium ion and a negatively charged chloride ion. Other chemicals that are frequently present as ions from dissolved salts include some important plant nutrients: calcium, sulfur as sulfate (SO_4^{-2}), nitrogen as nitrate (NO_3^{-1}), and phosphorus as phosphate (PO_4^{-3})

The level of dissolved salts is usually determined by running an electrical current through water and measuring the conductivity of the water over a given distance. Conductivity is the inverse of resistance. Ions are good conductors, whereas pure water is not; so the conductivity of water is a good way to measure the concentration of ions present. Conductivity of water is affected by temperature, so measurements must take that into account. Units of conductivity are measured in micro-Siemens per centimeter; a low level might be 50 μS/cm, and a high level, such as that of seawater, is 32,000 μS/cm. The Great Salt Lake of Utah registers an extreme value, 158,000 μS/cm. The level of dissolved salts in a lake is determined by the size and geology of the watershed; land use and human activities within the watershed (such as crop agriculture); atmospheric deposition; biological processes in the lake, particularly in the hypolimnion; and evaporation. Evaporation concentrates dissolved materials in the water left behind; this is why lakes in arid regions are often salt lakes, like the Great Salt Lake: evaporation rates are high in arid regions relative to rates of precipitation.

Closely related to dissolved salts is total dissolved solids. This is the total concentration of dissolved solids of all kinds in water. In natural waters dissolved salts are the main constituent, but others can be present including dissolved organic compounds, as well as toxic organic pollutants. While conductivity is often used to measure total dissolved solids, it really is measuring only the concentration of ions. A more accurate measure is taken by evaporating a water sample of known quantity and analyzing the solid residue left behind.

High levels of dissolved salts are a serious challenge to most aquatic organisms. Most living cells maintain structural integrity through internal pressure on the cell walls. In aquatic single- and multicelled organisms, this pressure is created by an osmotic differential between the contents of the cell, which have a higher ion concentration, and the surrounding water, which has a lower concentration. If such an organism is placed in salty water, water molecules will migrate across the cell wall by osmosis (toward the higher ion concentration) and the cell will dehydrate. Furthermore, inorganic ions migrate across the cell wall into the cell, where they are toxic above a certain concentration. Salt-loving one-celled organisms have adapted by maintaining higher internal concentrations of ions, though not necessarily the same ones as in the saltwater outside the cell. Potassium ions seem to be the preferred weapon to keep sodium ions at bay. Such organisms are termed halophilic. Some can tolerate high salt concentrations (facultative halophilic organisms), and some require high salt concentrations (obligate halophilic organisms).

Halophilic plants, or halophytes, use similar adaptations at the cellular level to live in saline conditions. Many, in addition, have specialized cells or organs that excrete salts to prevent their buildup as well as others that prevent salts from penetrating far into the plant, particularly the roots in the case of emergent vegetation. Animals that live in saltwater generally have adaptations that involve controlling the ion concentration of their bodies. Simple animals such as marine zooplankton maintain an internal osmotic concentration close to that of the surrounding water. Larger, more complex animals have specialized regulatory organs that perform as the kidneys do in humans, regulating salt concentration and removing excess salts for excretion. At the extremely high levels of salinity found in salt lakes, however, few organisms can survive. One exception is the brine shrimp, which are crustaceans of the genus *Artemia*. These small creatures can live in varying levels of salinity, including the very high. In hypersaline environments, a lack of competitors and aquatic predators can give rise to large populations of brine shrimp.

Interestingly, saltwater and freshwater fishes maintain about the same internal salinity levels. Saltwater is about three times as salty as their blood, so they must use specialized organs to collect, transport, and excrete salts that are constantly

Acidity and Alkalinity

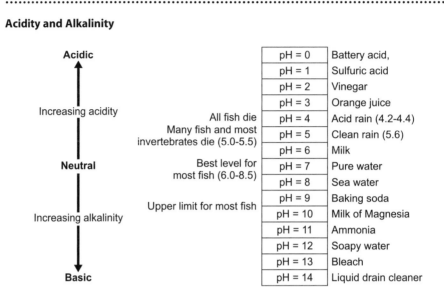

Figure 1.4 The pH scale. *(Illustration by Jeff Dixon. Adapted from U.S. Environmental Protection Agency, Acid Rain Students' Web site.)*

Acidity and alkalinity are used to describe the ion balance of water; the degree of acidity or alkalinity of water is indicated by its pH value. The pH scale (see Figure 1.4) ranges from 0 to 14. pH is the negative exponent of the concentration of free hydrogen ions (H^+) and reflects the balance of hydrogen ions and hydroxide ions (OH^-). If these are in balance, the pH is equal to 7, and the water is considered neutral. If there is a predominance of hydroxide ions, water is considered alkaline (pH >7 up to 14). If hydrogen ions predominate, pH will be less than 7, indicating acidic water. Because pH reflects the value of an exponent, it is a logarithmic scale. Each integer on the scale indicates a difference of a factor of 10 in the concentration of hydrogen ions.

"invading" because of osmotic pressure. Freshwater fish have the opposite problem: their higher internal salinity means that the water in which they live constantly threatens to "waterlog" them and must be collected and excreted.

pH is of great importance to many biological processes. The degree of solubility of dissolved substances, such as metals and nutrients, as well as organic compounds, is a function of pH. Given that the availability of nutrients such as phosphorus to organisms is dependent on their chemical form, and the chemical form is determined by pH, it is clear that pH is a critical factor in determining what types of organisms can live in a body of water, and what their population levels will be. For example, the bioavailability of calcium, necessary for formation of bones and shells, is reduced in acidic waters. In contrast, the bioavailability of metals is enhanced by increasing acidity (decreasing pH), which may result in toxicity.

In lakes, fish are usually the greatest management concern, and a pH range of 6–9 is optimal for most fish. Various fish species have different ranges of tolerance for both acidity and alkalinity, and within a species, different tolerances exist at different life stages. Tolerable pH for survival may be different than tolerable pH for reproduction. Acidification of freshwaters due to acid precipitation ("acid rain") and the effects of mining has impacted aquatic organisms and biotic communities.

The pH of lakes is variable and depends on a number of factors. One is the geologic composition of the watershed. Rocks such as limestone and dolomite remove hydrogen ions from water when they dissolve, resulting in water that is alkaline. Lakes fed by water from streams and groundwater running over and through such rock tend to be alkaline. Such lakes will be able to resist acidification in the face of acidic precipitation, both natural and polluted. Natural rainfall is slightly acidic (pH around 5.6) because atmospheric carbon dioxide in water forms carbonic acid. Air pollution from burning fossil fuels can dramatically increase the acidity and decrease the pH of precipitation; pH values as low as 3 have been recorded in some regions, and in those regions, acidification of lakes is often seen. High levels of dissolved organic carbon from vegetation acidifies some lakes, while pollutant discharge from industries may alter pH in others.

Biological processes can also alter pH. Photosynthesis uses carbon dioxide (CO_2) from water, thus increasing pH; respiration releases CO_2 into water, resulting in lower pH and greater acidity. In eutrophic lakes, pH can fluctuate daily.

Gases, too, dissolve in water, and surface waters in contact with the atmosphere will have dissolved gases largely but not precisely reflecting the chemical composition of the atmosphere. The atmosphere is about 78 percent molecular nitrogen gas (N_2) and 21 percent molecular oxygen gas (O_2); other atmospheric gases present in small concentrations include argon and carbon dioxide, as well as a number of "trace" gases. While dissolved molecular nitrogen has little biological importance in aquatic systems, oxygen has a great deal of importance.

Most living organisms require oxygen to release the energy stored in the carbon-based molecules that make up their food. This process is known as respiration, and chemically, it is the reverse of photosynthesis. In photosynthesis, plants use solar energy to combine carbon dioxide and water to produce carbohydrates. In respiration, plants, animals, and other organisms use oxygen to break down carbohydrates into carbon dioxide and water, releasing energy.

The ability of gases, including oxygen, to dissolve in water is an inverse function of temperature: the higher the temperature, the lower the maximum possible concentration of dissolved gases; the lower the temperature, the higher the concentration. If the dissolved oxygen concentration, measured in milligrams of oxygen per liter of water (mg/L), is at its maximum level for a given temperature, it is said to be at its *saturation level*. Liquid water just above freezing at sea level has the highest saturation level of oxygen at about 14.6 mg/L; this level decreases linearly with temperature. Elevation and barometric pressure also affect the saturation level:

atmospheric pressure decreases with elevation, so the saturation level at any given temperature also decreases.

The other influence on oxygen levels in water is biological activity. Plants, large or small, produce oxygen during photosynthesis (and use carbon dioxide, which is also dissolved in water). When photosynthesis takes place in water, oxygen levels increase. Microorganisms, plants, and animals all use oxygen in the process of respiration; this process diminishes oxygen levels.

In the near-surface zone of lakes, oxygen depleted by biological activity is renewed by diffusion from the atmosphere. But oxygen does not diffuse rapidly in water, so physical mixing of the water column is necessary to move the oxygen down into deeper water. The turbulence of rivers ensures that there is almost always a good supply of oxygen in them, but this is not the case in lakes. In the absence of active mixing, and particularly during periods of thermal stratification, oxygen levels can become severely depleted, especially in the dark depths where photosynthesis does not occur but where considerable decomposition does. Where photosynthesis does occur, oxygen levels may cycle up and down daily, and organisms have developed various strategies for dealing with these cycles.

All but a few freshwater aquatic organisms need oxygen to survive. But, compared with the atmosphere, oxygen levels in water are relatively low under the best of circumstances. For most fish species, optimal dissolved oxygen levels are between 7 and 9 mg/L. Cold, turbulent streams of the type favored by trout may have oxygen levels of 9–12 mg/L, which is near the upper limit found in freshwater environments. Under less-favorable circumstances, oxygen levels may become low; then fish and other organisms either move to where levels are higher, or die.

Aquatic organisms need to get oxygen into their bodies. Many use integumental respiration. This means they absorb oxygen through the body surface (skin or cell walls) and have no specialized organs of respiration. This is a good strategy for small organisms like phytoplankton and zooplankton, because the smaller the organism, the greater the ratio of body surface to body volume. And if an organism is very small, it has no need for a specialized system to distribute oxygen, which simply diffuses through the cells.

Such a passive way of getting oxygen has a drawback. Oxygen dissolves in water, but it does not diffuse through water as efficiently as through air. Therefore an aquatic organism, through its respiration, is liable to deplete the oxygen nearby. A layer of low-oxygen water will surround the organism.

Aquatic organisms have developed a number of ways to keep water moving and prevent a depleted layer of water from developing. Multicellular (though still small) organisms using integumental respiration solve this problem by moving about or simply flexing their bodies. Larger aquatic organisms, like fish, have much more sophisticated ways of dealing with the need for oxygen. Some contact the surface periodically to use atmospheric air. Some arthropods trap or store bubbles of air outside their bodies, like little scuba divers. Pulmonate or "lunged" snails, descended from terrestrial snails, store air in a cavity under their shell; this allows them to live in oxygen-depleted environments, such as wetlands. Some insects have snorkel-like appendages that enable them to acquire atmospheric air without emerging from the water.

Other organisms use the dissolved oxygen in water. Fishes have elaborate gill structures to increase vastly the surface area across which oxygen will diffuse; they "turbocharge" the rate of diffusion by keeping water moving across the gills, either by flapping a part of their body, or by moving forward, or both. Many aquatic larvae of insects and other macroinvertebrates use gill structures as well.

Lifeforms of Freshwater Aquatic Environments

Viruses

Viruses are extremely small (30–300 microns) and lack a cellular structure, but they are alive in the sense that they self-replicate and evolve. Without a cellular structure of their own, they can live only as parasites inside the cells of other organisms. They are ubiquitous, and many cause diseases in their host organisms.

Bacteria

Bacteria inhabit freshwater systems in large numbers, especially on organic detritus—which they help to decompose—in sediments, and in a biofilm covering all submerged surfaces. These surfaces include rock and gravel particles on the river or lake bottom, the skin of fishes, fragments of large woody debris, and sides of boats. Their concentrations in the water column are lower. Most are consumers of dissolved organic material (DOM). The autotrophic cyanobacteria, or blue-green algae, are a major planktonic group, particularly in nutrient-rich waters.

Fungi

Numerous fungi occur in freshwater environments, but the hyphomycete fungi are the most important. These tiny organisms colonize dead leaves, wood, and other organic detritus in streams, softening the tougher components and rendering them more palatable and more valuable as food to invertebrates. Their astronomically numerous spores are also consumed by specialized detritivores.

Algae

Algae are single- or multicelled organisms, mostly microscopic plants. Along with the cyanobacteria, they are the primary autotrophs (organisms that produce carbohydrates through photosynthesis) in freshwater systems. The algae may be grouped according to where they occur. Periphyton, or periphytic algae, are attached to substrate (rocks, sediments, or the surfaces of aquatic macrophytes); phytoplankton, or planktonic algae, are suspended in the water column.

Most periphyton, the plant component of biofilms, are composed of diatoms (class *Bacillariophyceae*). These beautiful lifeforms (see Plate I) consist of a single cell inside a glass (silica) shell. The shells (frustules) of different species of diatoms have unique shapes and patterns; a commonality is that they are all constructed of two halves that fit together like a hat box and lid. Some authors (for example, Allan 2001) include in the periphyton green algae and cyanobacteria.

Phytoplankters are, by strict definition, single-celled microscopic or near-microscopic plants suspended in the water column (see Plate I). However, other members of the plankton include protists and cyanobacteria. Many phytoplankton are periphyton that have become detached, particularly in fast-moving currents.

Macrophytes

Macrophytes are larger, multicellular plants that play a key role in many aquatic environments. In terms of growthform, macrophytes can be emergent, floating-leaved, free-floating, or submerged. Emergent plants are rooted in the underwater substrate, but they grow up out of the water. Floating-leaved plants are similarly rooted; their leaves float on the surface but do not extend further. Free-floating plants are not rooted but float freely on the surface and often form large floating mats upon which other forms of life become established (as in many tropical freshwater environments). Submerged plants are rooted but completely underwater.

Major aquatic macrophytes include bryophytes (mosses and liverworts, found typically in faster-moving, shallow waters), angiosperms (flowering plants), and some algae species large enough to be seen without a microscope. The abundance and distribution of macrophytes in freshwater aquatic environments are controlled by several factors. Light availability limits growth of some macrophytes in shaded streams and restricts populations to shallow margins of many larger rivers and lakes because of depth and frequent turbidity. Most aquatic macrophytes are not well adapted to high water velocities and thus are limited to backwaters and channel margins of rivers by current and to sheltered bays of lakes by wave action. Substrate conditions are often not suitable for the establishment of macrophytes, and nutrient availability may be limiting in nutrient-poor aquatic environments.

Protozoans

The protozoans are a diverse group of unicellular animals such as ciliates (protozoans with hairlike structures). Most are microscopic, though some can be seen with the unaided eye. They tend to prefer slackwater areas, depositional areas (areas where suspended material is deposited), and interstitial spaces in the substrate (that is, the spaces between particles of sand or gravel). Some protozoans graze on bacteria and algae; others are predators; still others are parasites. Protozoans in turn are consumed by many small invertebrates, including midge larvae.

Rotifers

Barely visible to the unaided eye, these tiny animals are an important and abundant constituent of the zooplankton (along with protozoans) in most freshwater environments (see Plate I). They constitute a major food source for some fish. The rotifers are a large group that consumes algae, bacteria, and smaller animals.

Flatworms

In streams and rivers, planaria (Order Tricladida) are the most important representatives of this group. Many prefer cold water and so are found mostly in headwater streams. Mostly flat or ribbon-like worms (0.2–1.2 in or 5–30 mm) long, planaria glide over the substrate, scavenging or hunting for prey.

Nematodes

Nematodes are unsegmented microscopic or near-microscopic roundworms. An extremely diverse and ubiquitous group, they inhabit marine, terrestrial, and freshwater aquatic environments. Many are parasitic, but some free-living species inhabit freshwater environments, where they are part of the microbial loop.

Annelid Worms

Of this large phylum, two main groups occur in freshwaters: *oligochaetes* and *hirudinae* (leeches). Most oligochetes are detritivores (scavengers in sediment deposits, for example, in the substrate of pools). Some leeches are scavengers; others are parasites. Annelids tend to tolerate low dissolved oxygen levels; if they are dominant, it is an indication of poor water quality.

Sponges

While the members of this phylum (Porifera) are usually associated with marine environments, about 150 sponges occur in freshwater environments throughout the world. Morphologically simple creatures, they range in size from less than 1 in (1–2 cm) to about 3 ft (1 m), and are typically brown, greenish, or yellow, according to the colors of the algae living on them. They attach to pieces of wood or other relatively stable pieces of substrate, and live by filtering algae, protozoans, and fine particles.

Molluscs

Two classes of molluscs are common in freshwaters: gastropods (snails and limpets) and pelecypods (the bivalve clams and mussels). Snails feed on periphyton and sometimes on detritus; they are herbivorous or omnivorous. Most freshwater bivalves are filter feeders, filtering fine particles of algae, bacteria, and organic detritus from the water. As such they play an important ecological role in reducing turbidity, particularly in lake environments. In turn molluscs are prey for fish, some bird species, crayfish, turtles, and mammals such as raccoons, muskrats, mink, and otters.

Crustaceans

Of the approximately 40,000 known species of crustaceans, only about 4,000 live in freshwater. The larger freshwater inhabitants include isopods (sowbugs), amphipods, decapods (crayfish and shrimp), and a few crabs. Freshwater aquatic isopods and amphipods inhabit clean, cold waters and are omnivorous scavengers. The decapods are detritivores, herbivores, and predators, some changing feeding habits with different life stages. All are relatively secretive inhabitants of the benthos; some isopods and amphipods inhabit the hyporheic zone as well. In turn these organisms are prey to a wide variety of predators, including fish, birds, snakes, and mammals.

Several taxa of microcrustaceans play important roles in food webs of freshwater ecosystems (see Plate I). These include members of the classes *Ostracoda* (sometimes known as seed shrimp or mussel shrimp), *Branchiopoda* (fairy shrimp, daphnia), and *Maxillopoda* (including the fantastically numerous copepods). They inhabit a variety of habitats within the stream environment, mainly in areas not exposed to fast currents, and feed on organic particles and one-celled plants and animals.

Insects

The vast majority of species of this enormous taxon are terrestrial, but aquatic insects account for the majority of macroinvertebrates in most rivers and streams. Most do not spend their entire lives in water, only their larval stages. For some, however, the larval stage is long compared with a relatively brief adult period. The major insect taxa that play important roles in freshwater systems are introduced below.

Mayflies (Ephemeroptera). The mayflies include more than 300 genera and more than 2,000 species and are found on every continent except Antarctica. Most species are obligate river-dwellers; only a few inhabit lakes and wetlands. The mayflies are known (and named) for their short adult lives, which may last from a few hours to a few days. Adults do not eat, but rather fly about looking for mates. After successfully finding one and mating, they die. The aquatic larvae are mostly herbivores and scavengers; few are predatory. Mayfly species can be found in all of the freshwater aquatic habitats and are an important food source for fish.

Caddisflies (Trichoptera). The caddisflies are a large insect order with almost 10,000 species described, and estimates of several times this many existing globally. They are moth-like, and unlike many other orders of aquatic insects, all are aquatic in their larval stage. The adults emerge, sometimes en masse, to mate and die. Caddisflies occupy the full range of freshwater habitats and have evolved a variety of ecological specializations. Some are predators; some spin nets to collect fine particulate organic material (FPOM); some graze, cow-like, on the biofilm; some shred plant material; and a few reportedly tend gardens of protozoa and algae. All can produce silk and most species use it to create cases that they inhabit like hermit crabs, often incorporating bits of organic detritus and small sediment particles into them. They are a major food source for fish and are widely distributed.

Stoneflies (Plecoptera). The stoneflies, as the common name suggests, tend to inhabit rocky or gravelly substrate. This implies headwater streams, with their generally high levels of oxygen and cool temperatures, and these are in fact the conditions that most stoneflies require. The nearly 2,000 species tend to be concentrated in the mid- and high-latitudes. The lives of adults are short compared with the larval stage. Adults, notoriously poor fliers, emerge and spend their lives near but not in water; some feed as adults but others have no mouthparts. Only one stonefly, an inhabitant of Lake Tahoe in the United States, is aquatic throughout its life.

Stoneflies exhibit the full gamut of feeding preferences: some are detritivores whose life cycles are timed to take advantage of the autumn pulse of leaves entering the stream; others are predators, consuming midge and blackfly larvae. Some species begin life as detritivores, but switch to predation as they grow; others eat whatever they can find. The larvae, eggs, and adults are consumed by a variety of fish species, birds, amphibians, and larger invertebrates such as hellgrammites.

True flies (Diptera). Although many flying insects are termed "fly" (for example, dragonfly, mayfly, caddisfly), only those belonging to the order Diptera are "true" flies. This order contains tens of thousands of species, including some truly annoying ones—mosquitos, black flies, midges, horse flies, and deer flies. Many are aquatic in the larval stage and a few occur only in running water (lotic) environments (for example, black fly larvae). Some of the important families in freshwater environments include black flies, with 1,650 known species; midges, with about 20,000 species known; and crane flies, whose known species number about 15,000. The larval black flies are almost all filter feeders, occur sometimes in high densities in the aquatic environment, and (especially in such high densities) are an important prey for a broad spectrum of fish, birds, and larger invertebrates.

Midges are broadly distributed and occupy the full range of aquatic habitats. Most are collector-gatherers feeding on FPOM. Some wield net-like body parts to gather food particles; others graze on the biofilm. Still others burrow into macrophytes or consume wood. A few are predators. The midges, or chironomids, are themselves consumed in huge numbers by larger invertebrates, fish, amphibians, and birds.

The crane flies likewise occupy a broad variety of habitats, many lotic. They are primarily detritivores and herbivores, but some are predators.

Beetles (Coleoptera). The Web site of the Coleopterists Society contains the following description of beetles:

> Beetles ... are the dominant form of life on earth. One of every five living species of all animals or plants is a beetle! Various species live in nearly every habitat except the open sea, and for every possible kind of food, there's probably at least one beetle species that eats it.

The sheer number of species of beetles gave rise to the biologist J. S. Haldane's famous comment that one could infer from the study of nature that the Creator has "an inordinate fondness for beetles." More than 300,000 species have been described, and it is estimated that the total number is many times that. Beetles live in every terrestrial environment except Antarctica, so it should not be surprising that some (but not many) are aquatic. Some familiar to those who study aquatic environments include the water-penny, the riffle-beetle, diving beetles, and whirligig beetles. Aquatic beetles may be detritivores, grazers, or predators, with some changing feeding behavior at different life stages.

True bugs (Hemiptera). This large order has more than 3,300 species living in or on water. The most conspicuous is probably the giant water bug (family Belostomatidae) or "toe-biter," which as its name suggests can inflict a painful bite. It has powerful pinchers because it is a predator, as are many of its fellow Hemipterans. Few are large enough to attack fishes, frogs, or water snakes as the giant water bug reportedly does. Other familiar members of this order include water striders (Gerromorpha), backswimmers and water crickets (Nepomorpha).

Alderflies, dobsonflies, and fishflies (Megaloptera). This medium-size, mostly terrestrial order contains some of the freshwater environment's best-known aquatic insects, in particular the spectacularly large and voracious dobsonfly larva, also known as the hellgrammite. Hellgrammites are an aquatic insect that one should handle with care; the author has personally witnessed a large specimen pinching hard enough with its powerful jaws to draw blood. In their early larval stages, most of the Megaloptera tend to be detritivores; however, in their later larval stages, they become predators, consuming a range of invertebrates. Alderflies and fishflies share many of the characteristics of their larger cousins.

Dragonflies and damselflies (Odonata). The adult members of this order are known even to the least observant frequenter of freshwater environments. Most people see them only in their winged adult forms, which include many large dragonflies and gorgeously colored damselflies. All are efficient predators in their aquatic, larval stage and their adult stage, first consuming aquatic invertebrates and small vertebrates and later catching and eating flying insects on the wing. They are consumed by fish attracted by the generally large size of the larvae.

Vertebrates

Fishes. The fishes are the dominant aquatic lifeform both in terms of their importance in the freshwater food web and their interest to humans. All fishes are obligate aquatics. Although other vertebrates live in and around freshwater environments and are part of the aquatic food web, relatively few spend their entire lives in the water.

The fishes are a vast group; about 25,000 species are scientifically described. Of these, about 41 percent or more than 10,000 are freshwater and 1 percent move between freshwater and the sea during their lives. The fishes have adapted to conditions in practically every aquatic environment on Earth. They range in size from the very small (less than 0.5 in or about 1 cm long) to the very large (the whale shark can reach 39 feet or 12 meters in length). They may be shaped like discs, pencils, boxes, doormats, and basketballs. Some fish have scales and others do not; some fish are brilliantly colorful, others are drab. Some eat plants (herbivores), some eat invertebrates, some eat detritus, some eat other fish, and some (in a pinch) will eat people.

The vast majority of fishes, including the most numerous and important freshwater ones, are teleosts, members of the division Teleostei. The most important subdivision—both in terms of numbers of species and in terms of numbers of important freshwater species—are the Euteleostei, or true teleosts.

Other freshwater fishes are only remotely related to the teleosts. For example, some lampreys inhabit freshwaters; they belong to the order Petromyzontiformes, a group characterized by an absence of jaws. Several representatives of the cartilaginous fishes, Order Rajiiformes, commonly known as skates, live in freshwater. Primitives—fishes that have survived in much the same form as their ancestors, which lived hundreds of millions of years ago—are present. These include several species of lungfish in the orders Ceratodontiformes and Lepidosireniformes, as well as ancient order Acipenseriformes, with 24 species of commercially valuable and much overfished sturgeons (family Acipenseridae) and the strange-looking paddlefish (family Polyodontidae, two species). Eleven other primitives include bichirs and reedfish (family Polypteridae) in the order Polypteriformes, and the predatory, air-breathing gars (seven species) and bowfins (one species) in the order Semionotiformes.

Compared with the teleosts, the fishes just named, though interesting, are small change. The teleosts include 38 orders comprising 426 families and 23,600 species. Of these, more than 22,000 species are true teleosts. Members of the superorder Ostariophysi (4,000 species) dominate freshwaters. This large group includes the order Cypriniformes (minnows, barbs, and carp), which contains the largest family of freshwater fishes, the cyprinids. The cyprinids, however, have no representatives in South America, where the characids seem to be their functional equivalent. An important tropical order is the Characiformes, which includes—among many others—the infamous piranhas that inhabit the Amazon basin and elsewhere. The catfishes (order Siluriformes) are surprisingly numerous and diverse.

Another large group (superorder) of true teleosts with several important freshwater families is the Protacanthopterygii, which includes the order Salmoniformes. Whitefishes, graylings, chars, trout, and salmon all belong to this order. The superorder Percomorpha contains more than 12,000 mostly marine species but also some important freshwater members. In fact, one order within this group, Perciformes, is not only the largest order of freshwater fishes—148 families and 9,300

species—but is the largest order of vertebrates of any kind. Within it are the centrarchids (family Centrarchidae), whose members include sunfishes, crappies, rock-
basses, and the black basses so beloved of North American anglers. The family
Percidae includes perches and darters, of which the greatest diversity is found in
the Tennessee River basin in the United States. The large and illustrious family
Cichlidae has more than 1,500 species and is important in the tropics, particularly
but not exclusively in Africa. In some of Africa's lakes, in which most of the species
are endemics, cichlids have demonstrated explosive speciation, with endless variations of size, shape, color, and behavior (mating, parenting, feeding strategies).

Amphibians. All amphibians need some form of freshwater habitat for at least a
part of their lives; many prefer lentic (still water) habitats. Members of three
orders—Anura, the frogs and toads; Gymnophiona, the caecilians; and Caudata,
salamanders and newts—live in lotic (flowing water) habitats at least for part of their
lives. A great many frogs and toads breed in streams or along the margins of freshwater bodies, and many remain closely tied to them for the entire lives. The caecilians are legless creatures; most live in soil and are capable of moving rapidly through
it, but a few species live in water. The Caudata contain a number of species, like
some dusky salamanders (genus *Desmognathus*), that are obligate aquatics—that is,
they must live in water their entire lives. However, even those that are terrestrial as
adults use freshwater environments for breeding. Some salamanders are important
predators in lotic systems by virtue of either relatively large size (for example, hell-
benders) or dense populations in low-order streams. Amphibians in aquatic systems,
lentic or lotic, tend to be grazers in their larval stage and predators as adults.

Reptiles. A number of reptiles, in particular, members of the order Crocodylia,
snakes (Squamata) of many families, and freshwater turtles (Testudines, sometimes
referred to as chelonians), play important roles in freshwater food webs. Crocodiles, alligators, and related species are voracious predators that may feed on invertebrates as well as fish, especially early in their lives. Some crocodiles and
alligators grow to impressive sizes, and it is likely that their impact on the fish population is substantial. They mostly inhabit lowland, sluggish streams and rivers,
lakes and ponds, and wetlands. Snakes are almost all predators, preying on fish primarily but also invertebrates. Terrapins tend to be omnivorous. The fearsome alligator snapping turtle is a good example, eating fish, invertebrates, and almost
anything else to grow to an average of 175 lb (80 kg).

Birds. Many birds live along rivers, in wetlands, and on and around lakes, and are
involved in the aquatic food web, in many cases as top carnivores (see Plate II).
Ducks, geese, and other waterfowl forage in shallows of lakes and rivers. Eagles,
ospreys, kingfishers, and various wading birds are fish-eaters (piscivores) but are
not limited to aquatic habitats, nor do they spend much time actually *in* the water.

The adult stages of aquatic insect larvae feed many birds that are not themselves in any sense aquatic.

Many species of birds, however, can be considered truly aquatic in the sense that they spend a good deal of their lives either on the water or underwater. Cormorants and anhingas, for example, are quite as much at home underwater, where they pursue fish, as in the air. Some water birds prefer moving water, even rapidly moving water. The dippers wade or swim underwater in rapids, typically in low-order mountain streams, looking for insect larvae and the occasional fish or crustacean. There are also several ducks that seek out invertebrates in fast-moving waters. A list of bird orders with member species that are associated with freshwater environments follows.

Gaviiformes. Loons or Divers. All are aquatic, seldom coming ashore. They inhabit lakes, wetlands, and rivers throughout the higher latitudes of the Northern Hemisphere. Using their powerful feet to propel themselves underwater, they feed on a variety of aquatic organisms, primarily fish, amphibians, and crustaceans.

Podicipediformes. About 20 species of grebe, as well as the New Zealand Dabchick. These birds, at home in the water, are ungainly on land, where they only go to nest. They are widely distributed in both the Northern and Southern Hemispheres.

Pelecaniformes. More than 50 waterbirds including the well-known pelican (*Pelicanus* spp.). Most are seabirds, but the cormorant (*Phalacrocorax* spp.) is a familiar freshwater diving bird that preys mostly on fish but also eats water snakes.

Ciconiiformes. Large wading birds with elongated legs and specialized bills, including storks, herons, egrets, ibises, and spoonbills. They stalk fishes, crustaceans, amphibians, and other prey in the shallow water on the margins of lakes and slow-moving rivers as well as in wetlands.

Phoenicopteriformes. Flamingos. This order has but one family, and one genus, and six species (*Phoenicopterus* spp.). These long-necked, brilliantly colored wading birds are widely distributed through Africa, southern Eurasia, and the Neotropics. While mostly found in saline or brackish environments, including salt lakes, they sometimes inhabit freshwater wetlands and lake margins in large numbers.

Anseriformes. Ducks, geese, and swans. The family Anatidae contains more than 140 species of these familiar waterfowl. They are widely distributed; many are powerful fliers and migrate long distances. All are well adapted to aquatic environments, where they float on the surface and dive in shallow waters to feed on a wide variety of plants and invertebrates. Some are piscivores.

Falconiformes. Hawks, eagles, and osprey. Members of two widespread families in this group of predators participate almost exclusively in aquatic food webs.

Accipitridae includes the fish eagles and America's Bald Eagle (*Haliaeetus leucocephalus*), whose diets are either exclusively or primarily fish. The Osprey (*Pandion haliaetus*) feeds almost exclusively on fish from both fresh- and saltwater environments.

Gruiformes. Several families are primarily or exclusively inhabitants of freshwater environments. The family Rallidae contains about 130 species of rails, coots, crakes, and flufftails. Typically, these birds are secretive, dwelling in dense vegetation (particularly reeds) in wetlands and along the margins of lakes and rivers, where they exploit the available food resources—small fish, snails, insects, and seeds.

Other aquatic or semiaquatic gruiformes include the tropical Sungrebe and Finfoot (two species), the Sunbittern, the crane-like Limpkin, and 15 species of cranes (family Gruidae), including the critically endangered Whooping Crane.

Charadriiformes. Shorebirds, gulls, and terns. While many members of this diverse order are limited to the seacoast, some live around freshwater habitats, especially wetlands. Woodcocks, for example, have an elongated bill designed for feeding on worms and burrowing invertebrates in swamps and bogs of the Northern Hemisphere. Sandpipers are often associated with seashores, but they also can be found along the shores of lakes, wetlands, and rivers.

Several charadriiformes are found in both marine coastal habitats and salt or soda lakes. The avocets (genus *Recurvirostra*) and stilts (such as the Black-necked Stilt) are examples. Their long slender bills, designed for probing into soft sandy sediments for invertebrates, and extremely elongated legs are as well adapted to the margins of salt lakes as they are to estuarine tidal flats.

Gulls (family Laridae) and their near relatives, the terns (family Sternidae), are also in this order. Often associated with marine environments, many of these birds also occupy freshwater environments, where some prey primarily on insects and others eat predominantly fish. In lakes such as Lake Ontario, the gulls' position at the top of the aquatic food chain has made them susceptible to the effects of bioaccumulated toxic chemicals.

Jacanas (family Jacanidae) are prominent in the Tropics throughout the world. Their unique ability to walk on lily pads and other floating vegetation is based on their enormous feet, which serve to distribute their weight. Their preferred habitat is the margins of shallow lakes and wetlands, where they consume insects and other invertebrates.

Coraciformes. Kingfishers. One of the most conspicuous inhabitants of streams and rivers and their banks in many parts of the world are the kingfishers. Although not all kingfishers are associated with freshwater environments, two families, Cerylidae (the Belted Kingfishes) and Alcedinidae (the River Kingfishers) are. Most kingfishers are in the latter family and inhabit Eurasia and Australasia. All American Kingfishers, including the noisy and conspicuous Belted Kingfisher, are in the former family. These flashy birds are often seen perched on branches overhanging the margins of rivers, from where they prey on small fishes.

Passeriformes. Perching birds or songbirds. Many passerines are connected to the aquatic food web, mostly by feeding on flying insects that spend their larval stages in freshwater. A few, however, are more closely associated with freshwater environments.

Mammals. Many mammals use riparian environments and are connected to aquatic food webs, but few actually inhabit the water. For example, brown bears feast on salmon returning upstream to spawn, but no one would call them aquatic organisms. Mammals that do spend a significant part of their lives in freshwater include otters (family Mustelidae, subfamily Lutrinae) and a few other mustelids, beavers (family Castoridae), hippopotami (family Hippopotamidae), river dolphins (family Platanistoidae), and the Australian platypus (family Ornithorhinchidae). Some, like the otters, are predators, feeding on fish and invertebrates; others, like the hippos, graze aquatic plants.

Appendix

Selected Animals of the Freshwater Aquatic Biome

Freshwater Aquatic Biome

Invertebrates

Stonefly (true aquatic)	*Capnia lacustra*
Black flies	Family Simuliidae
Midges	Family Chronomidae
Crane flies	Family Tipulidae

Amphibian

Hellbender	*Cryptobranchus alleganiensis*

Reptile

Alligator snapping turtle	*Macrochelys temminckii*

Fish

Upper Klamath sucker	*Chasmistes brevirostris*

Birds

New Zealand Dabchick	*Poliocephalus rufopectus*
Bald Eagle	*Haliaeetus leucocephalus*
Osprey	*Pandion haliaetus*
Sungrebe	*Heliornis fulica*
Sunbittern	*Eurypyga helias*
Limpkin	*Aramus guarauna*
Whooping Crane	*Grus Americana*
Belted Kingfisher	*Megaceryle alcyon*

Mammal

Brown bear	*Ursus arctos*

2

Rivers

Introduction

Rivers occur in nearly every terrestrial geographic setting on Earth. Even in the frozen landscapes of Antarctica there are two rivers. Every terrestrial biome, from tropical rainforests to deserts, includes rivers as part of its mosaic of habitats. This diversity of terrestrial and climate settings, from rainforest to desert, influences every aspect of the river environment and results in what has been referred to as a "bewildering variety of natural features and degrees of human impact"—and this was only talking about the rivers of North America.

A river and its stream network make up part of the freshwater aquatic biome. A river is a large natural flowing body of water; a river system includes the mainstem of the river and its tributaries, which if they are relatively small are termed streams. To refer to a single or distinct "river biome" troubles some. Wetlands that are part of a river system can be viewed as part of the same biome, as can wetlands that exist at the margins of lakes. The dividing lines are not sharp. However, important differences do exist and treating these three freshwater habitats separately has the virtue of being consistent with much of the published literature. This chapter explores the tremendous variety of conditions—in vegetation, animal life including life histories, and environmental conditions—within river systems. In addition to describing variations within river systems, this chapter also explores the differences in these characteristics among rivers in different terrestrial biomes by describing rivers from three different terrestrial settings.

Human impacts on rivers are so widespread that today "pristine" rivers are rare in the world. Pristine rivers are not dammed; are not channelized; are not affected by agriculture, mining, logging, or urban development in their watersheds; and are unmolested by the introduction of nonnative species. Few rivers are in such a condition, and as the human population climbs from its current 6 billion plus to an expected 9 billion by mid-century, and as economies grow throughout the world, the prospects for those that are relatively pristine are not good. Suffice it to say that the biota of freshwater systems, taken as a whole, is subject to greater human impacts than those of almost any other biome.

The River Environment

Rivers and streams constitute the most dynamic of the freshwater aquatic environments. Their defining characteristic, flowing water, keeps the physical conditions of life constantly changing. Flowing water shapes and reshapes river channels, streamside (riparian) areas, floodplains, riffles and pools, and all the other physical parameters associated with river systems. The volume of flow also changes dynamically, with four to five orders of magnitude variation not uncommon. For example, the James River at Richmond, Virginia, had one occasion in recorded history (a short period in the life of a river) during which the discharge was observed to be zero or close to it. But in 1972, the peak discharge—same river, same place—resulting from a tropical storm system was well over 300,000 ft^3/sec.

For an aquatic organism such as a fish or macroinvertebrate, the dynamism of the moving water creates both opportunity and danger. As the rate of flow rises and falls, velocity increases and decreases. Too much velocity, and some organisms risk being swept away; too little, and those that depend on the current to bring them food will go hungry. To better appreciate the adaptations of organisms to life in moving water, it is necessary to understand the lotic environment—that is, the moving-water environment.

Discharge, or streamflow, is the volume of water passing a particular point (or more precisely, passing through a particular cross-section) on a stream or river per unit of time, typically per second. Volume may be reported in either cubic feet or cubic meters. Discharge, therefore, can be given in units of cubic feet per second (ft^3/sec), typically written as cfs, or cubic meters per second (m^3/sec), sometimes called cumecs.

Natural stream channels vary greatly in physical form. They are deeper in some places, shallower in others; wider in some places, narrower in others. This poses a challenge for the accurate measurement of discharge. For example, velocity is greatest near the center of the channel and slower near the bottom and sides. Nonetheless, by taking multiple measurements and averaging, hydrologists are able to measure discharge with some precision.

Velocity varies over time and, at any given moment, upstream or downstream. Variation in velocity over time at a single point on a stream results from changes in discharge. More water being fed into a stream—for example, as the result of a sudden downpour—increases width, depth, and velocity; as the discharge later diminishes, width, depth, and velocity all diminish too in most natural stream channels. Thus, the amount and quality of habitat available to stream organisms changes as well, expanding and contracting.

At constant discharge, velocity may be different at different points on the stream. Three variables affect velocity: slope, channel irregularities, and viscosity. Slope of any stream or river reach is the ratio of vertical drop to horizontal distance: the steepness of the channel. It can be represented as a percent or in terms of the angle. Sometimes the slope of a river is represented in terms of feet per mile, meaning elevation drop in feet per horizontal mile of river. A very steep reach of the Upper Youghiegheny River in the United States drops more than 200 ft/mi, making it a favorite among advanced whitewater boaters. The Amazon, on the other hand, drops only 0.053 ft/mi over about 2,500 mi (1 cm/km over 4,000 km) once it leaves the Andes, which is why its flooding is so widespread during the wet season.

The energy available to water in a river is provided by the force of gravity, and that force acts perpendicular to the Earth's surface. Thus, the steeper the slope, the greater the acceleration due to gravity's pull. This energy is what accelerates water to its observed velocity. In natural rivers, the vast majority of the energy of moving water is transformed into heat through turbulence, one reason steep mountain streams resist freezing solid.

Channel irregularities are any characteristic of the river channel that impede flow and cause turbulence. Variations in width and depth count here, as does channel roughness. Roughness can result from irregularities in the channel profile, such as a drop-pool pattern; from rocks, logs, or other "roughness elements," such as cars and washing machines in some abused Appalachian rivers; and from channel meanderings, which cause turbulence and dissipate the energy that otherwise would speed the water on.

Energy is also used in entraining (picking up) and transporting sediment, whether grains of sand or boulders. Sediment load—the amount of suspended sand, silt, gravel, and other materials being carried by the stream—affects velocity. The greater the sediment load, the lower the velocity.

Finally, viscosity is a determinant of velocity. In water, viscosity varies with temperature, with cooler water more resistant to flow. In most cases, viscosity has a relatively minor influence on velocity as compared with slope, roughness, and sediment load.

The dynamic and varied conditions under which river plants and animals evolved and survive are related in part to the motion of water, and in part to the physical habitat conditions that result from that motion. The physical form of rivers mirrors the rich variety of habitats created by running water.

Following are the major determinants of river habitat type and quality:

- Size of watershed
- Terrain
- Underlying geology of watershed

- Soils of watershed
- Climate
- Terrestrial vegetation
- Discharge amount, timing, and rate of change (flow regime)
- Water quality parameters
- Temperature (average and range of variability)
- Dissolved solids levels
- Dissolved oxygen levels
- Nutrient availability, particularly of nitrogen (N) and phosphorus (P)
- Turbidity
- Presence and characteristics of pollutants including fine sediments
- Channel substrate, cover, and form
- Condition of the riparian corridor
- Disturbance regime and disturbance history
- Introduced species of plants and animals

River Channel Form

River channels provide the physical setting for lotic communities. The variety of materials that comprise them and flow conditions within them provide microhabitats for river organisms.

Rivers adjust channel form—channel slope, size, and shape—in response to changes in sediment load (its total mass and average particle size) and flow regime (average discharge, peak discharge, and seasonal fluctuations). The size and shape of river channels represent a balancing of discharge and sediment load. In a stable climate, and in the absence of significant tectonic activity, a river's channel tends toward an equilibrium form that is "just right" for its usual flow regime and sediment load. River channels respond in the short term to weather events (tropical storms, for example), and extreme events can drastically alter channel form. But over time, even a river channel "blown out" by a flood will seek to return to what Luna Leopold called its "most probable state"—the channel form that the river is most likely to take at any point in time. This is its equilibrium form.

Geomorphologists have learned that this "most probable" form is determined by a storm event of an intensity that is reached, on average, once every 1.5 to 2 years. Larger floods will affect the channel, but the bigger the flood, the more seldom it occurs. The 1.5- to 2-year storm creates enough flow to move sediment and reshape the channel, *and* it occurs relatively often. Any lesser flow than that resulting from the 1.5- to 2-year storm will not shape the channel, as it lacks sufficient flow and energy to move channel materials.

The form of stream or river channels is described in three ways: longitudinally (headwaters to mouth or confluence), in planform (looking down from above), and in cross-section.

Longitudinal attributes. The elevation profile of the "typical" river is concave; in the headwaters, slopes tend to be steeper, while near the mouth of the river, slopes

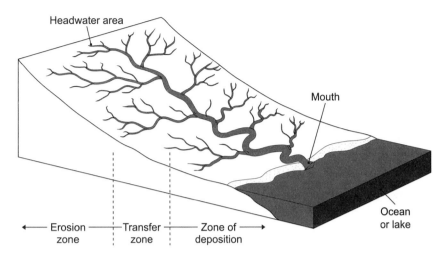

Figure 2.1 Concave longitudinal profile of typical river. *(Illustration by Jeff Dixon.)*

diminish toward the horizontal. River scientists find it useful to divide a river, considered longitudinally, into the zone of erosion, the zone of transport, and the zone of deposition (see Figure 2.1).

The zone of erosion is typically the headwaters of the stream or river and includes the upper part of the watershed and the small, first- and second-order streams that drain it. Headwater streams are typically steep, so flowing water has more energy to move sediment and larger sediment particles. These streams are actively cutting down, often through solid rock, and are likely to feature waterfalls and rapids. They are also likely to have water with relatively high dissolved oxygen levels and low temperatures, facts of great significance to the biota.

Headwater streams are typically characterized by channels whose bottom and sides are primarily bedrock. Because of the energy available in steep slopes for moving sediment, such streams often scour right down to the "living rock." Such stream channels are relatively straight (lack sinuosity) and have only the beginnings of floodplains, or none at all.

In the transfer zone, rates of deposition and erosion are approximately equal. This zone corresponds to mid-size (third- through sixth-order) streams. Sediment is moved through, with no net change in elevation or cross-sectional shape of the stream channel. Stream channels in the transfer zone have a mix of features characteristic of the headwater zone and the zone of deposition. Floating down such a river, one might pass over ledges and through rapids formed by relatively durable bedrock, into pools filled with sediment. On one side or the other, one might see well-developed floodplains, though not of great breadth.

In the zone of deposition, sediment is deposited and floodplains are built up. Even though water velocity may be as great or greater than headwater streams, due to lack of channel roughness, the ability of the river to carry sediment is diminished because of the low channel slope and hence much of it is deposited. To see exposed bedrock is unusual. The channel bottom and sides are carved from alluvium; such a channel is termed an alluvial channel. Where rivers meet the sea, and slope goes to zero, deltas may form. If sediment transport is interrupted, for example, through construction of levees or large dams and reservoirs upstream, delta regions can be starved of sediment and subside into the sea.

On a more local scale, certain regularities of the longitudinal profile of a stream or river channel are of great importance to the biota. In bedrock stream channels, a drop-pool sequence is often observed. In alluvial streams, a repeating sequence of pool and riffle (see Figure 2.2) is common. In riffles, the bed material is coarse—depending on the size of the stream, it could be composed of gravel, cobble, or even boulders. The slope of the channel is steeper than the average slope; depth is less. With more energy available to move sediment, finer sediments are absent and considerable space exists between the large sediment particles. These interstitial spaces, through which water can move readily, provide habitat for numerous species of small fish and macroinvertebrates. Sometimes riffles and pools are separated by runs, which are narrow, straight, relatively steep and fast stream sections. Runs are typically longer, narrower, and deeper than riffles.

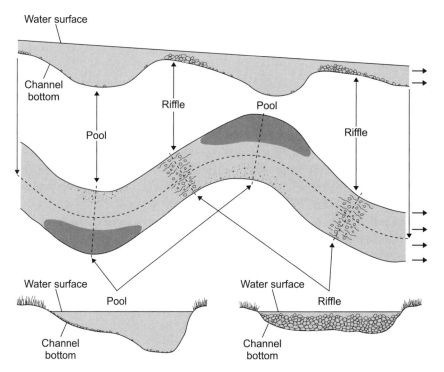

Figure 2.2 Plan view, longitudinal profile, and cross-sections of pool-riffle channel type. *(Illustration by Jeff Dixon. Adapted from U.S. EPA OWOW 1997.)*

In pools, the slope is lower than the average, and the water is deeper. Stream bed material is relatively fine, so the bottoms tend to be "mucky." Pools provide habitat for a different assemblage of fishes and macroinvertebrates than do riffles. The pool-riffle sequence provides habitat diversity that contributes to the overall biodiversity within stream systems. The pool-riffle-run sequence is associated with meanders in rivers, which are discussed below.

Planform characteristics of rivers. Looking down at a river reach from above, one of the most striking things is that rivers are not straight. This may be attributed, in part, to the fact that the land surface the river is cutting through is varied. Some parts are easily eroded, and others less so; rivers will follow the path of least resistance, and it is unlikely that path will be straight. But even in the absence of such heterogeneity, rivers do not run straight. They are sinuous, and even if forced by engineers into a straight channel, they will work to reestablish their natural curves.

The factors that make rivers tend toward sinuosity are complex. Slope is one of the channel attributes that rivers can adjust in response to changes in the flow regime or sediment load. For a given flow regime and sediment load, there is an equilibrium slope. Rivers adjust slope by adjusting sinuosity, much like what skiers, snowboarders, and skateboarders do when descending a hill. By weaving back and forth, taking a sinuous line, these athletes increase the length of their downhill run. Increasing this length while descending the same vertical distance decreases slope. Since decreasing slope means less energy being supplied per second, the skiers, snowboarders, and skateboarders do not accelerate as quickly. They control their speed, and have more fun too.

In a river or stream, energy, supplied by gravity, is used in eroding channel banks, transporting sediment, and in creating turbulence. So, although they are going downhill not uphill, rivers expend energy and do work. They adjust their rate of energy expenditure by adjusting slope, and they adjust slope by becoming more or less sinuous. In the process, they produce the characteristic curving form known as meandering, named for the river called Maeander by the ancient Greeks, now the Büyük Menderes River, in Turkey. In alluvial streams and rivers, meanders are often (but not always) associated with the pool-riffle sequence described above. Pools are found at the outside of the meander curves, and riffles on the "straights" between the meander curves.

The term "meander" is also a verb. Rivers meander, and research shows that every point on the floodplain of a meandering river has been occupied by the river channel at some time. Meandering, the river channel moves—but its cross-sectional shape (see next section) remains constant. The outside of the meander curve is an area of active erosion, and the inside of the curve is a depositional area. Thus, the channel over time will move toward the outside of the meander. Sometimes, in the process of meandering, rivers will double back on themselves, leaving meander curves cut off from the active channel. Oxbow lakes are thus formed.

Channel cross-section. The shape of the channel cross-section is the shape of the channel that would be observed if one sliced it with an imaginary plane perpendicular to the direction of flow. Many variations in channel cross-section are observed in natural rivers, both among natural rivers in different geographic settings and within the same river system (for example, headwaters versus depositional zone). Just as the pool-riffle sequence of variation in longitudinal profile is related to the meander pattern, so too are variations in cross-section. Pools characteristically form at the outside of meander bends. The channel at that point is relatively narrow and deep. Channel cross-sections where riffles occur tend to be relatively wide and shallow (see Figure 2.2).

In geographic settings in which erosion is rapid and slopes are steep, braided channels are likely to form (see Figure 2.3). In cross-section, they are much wider than they are deep. They also have low sinuosity. They appear choked with sediment and typically form under conditions of high sediment load. Instead of a single channel, multiple channels are separated by islands of sediment. Braided channels are ever-changing, with channels forming and reforming on a relatively short timescale. They are common in rivers that flow through sand and gravel and have easily eroded banks.

A stable stream has three visible features in the channel cross-section: a low flow channel or thalweg; a bankfull channel; and a flood channel. In an alluvial stream, these features are easily identifiable. The low flow channel is the deepest part of the stream channel in any cross-section, and connecting all such points along the river's course delineates a channel. During periods of low flow, this would be the only visible part of the channel in which flowing water would be

Figure 2.3 A braided channel river in Alaska's Arctic National Wildlife Refuge. *(Photo by E. Rhode, courtesy of U.S. Fish and Wildlife Service.)*

found (water is also likely to be moving though the sediment beneath the channel bed, but it would not be visible on the surface).

The bankfull channel in a stable alluvial stream is defined by where the floodplain begins. There is typically a break in slope, more or less dramatic, that defines the top of the bankfull channel. This break is sometimes identified as being where the tops of point bars meet the floodplain floor. The bankfull channel is the channel that is formed by, and just adequate to accommodate, the bankfull flow, which is the high flow event that occurs with a frequency of 1.5 to 2 years on average (see Figure 2.4).

Figure 2.4 Second-order stream in Montgomery County, Virginia, at bankfull (top) and at a more common low water level (bottom). *(Photos by author.)*

Finally, the flood channel is the channel that carries greater-than-bankfull flows. In an unaltered, stable alluvial stream, it corresponds with the floodplain. It will be inundated no more frequently than once every two to three years on average. Sometimes alluvial streams also have benches or terraces, which are relics of floodplains from past climates.

A less-apparent feature of the channel is the hyporheic zone. This lies under the channel bottom and along its sides where an exchange of water occurs between the channel and the groundwater system. Because it is not obvious to a riverside observer, its importance as habitat was once neglected. However, it appears to be an active ecotone that functions as an important refuge for benthic (bottom-dwelling) organisms, particularly during low water periods. The hyporheic zone is connected with the main channel by flows not merely of water but also of organic material and nutrients. It provides habitat for a number of species, as suggested by riverine macroinvertebrates found in wells on the floodplain of the Flathead River in Montana (United States) as much as 1.2 mi (2 km) from the main channel.

Watersheds and Subwatersheds

Every river drains a watershed. Large rivers drain large watersheds, and for such watersheds the term "river basin" is often used (see Table 2.1). A watershed is a bowl-shaped area that drains to a particular outlet (see Figure 2.5). In the case of a large river basin, the outlet would be the mouth of the river, where it meets the sea. In the case of a smaller tributary, the outlet point would be the confluence, or junction, with a larger river.

Rivers and their watersheds are created together in ongoing geomorphic processes of erosion, sediment transfer, and deposition. The rate at which the surface of the watershed is eroded is a function primarily of climate and geology.

Table 2.1 The 10 Largest River Basins in the World

RIVER	COUNTRY AT RIVER MOUTH	WATERSHED AREA (km^2)	WATERSHED AREA (miles^2)
Amazon	Brazil	7,180,000	2,772,213
Congo	Democratic Republic of Congo/Angola	3,822,000	1,475,682
Mississippi	United States	3,221,000	1,243,635
Ob-Irtysh	Russian Federation	2,975,000	1,148,654
Nile	Egypt, Arab Republic of	2,881,000	1,112,360
Rio de la Plata	Argentina/Uruguay	2,650,000	1,023,171
Yenisei	Russian Federation	2,605,000	1,005,796
Lena	Russian Federation	2,490,000	961,394
Niger	Nigeria	2,092,000	807,726
Yangtze	China	1,970,000	760,621

Source: Czaya 1981.

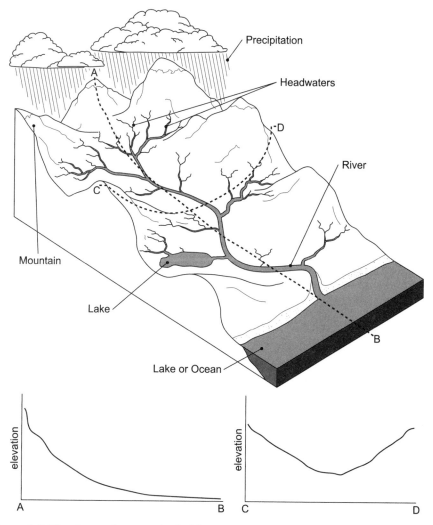

Figure 2.5 The shape of a watershed. *(Illustration by Jeff Dixon.)*

All but the very largest basins are subwatersheds—that is, they are part of a still larger watershed, and the rivers that drain them are tributaries of the larger river. Subwatersheds can be very large—the Ohio River basin, for example, is a subwatershed of the Mississippi. One way of understanding the relative size and position of a subwatershed in a larger basin is by the use of river order. For any tributary stream in a river system, a stream order number can be assigned. The most widely used stream ordering system is that developed in 1952 by Arthur Strahler. His system is simple: all headwaters streams with no tributaries are given the order 1. When two first-order streams come together, the resulting stream is a second-order stream. If two second-order streams join together, the downstream segment is third

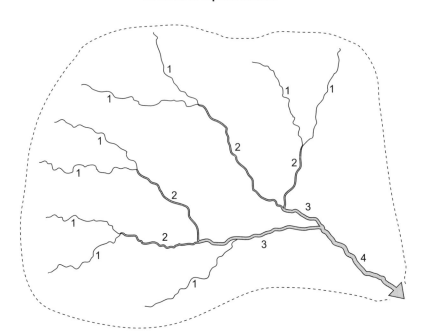

Figure 2.6 Stream ordering according to Strahler (1952). First-order streams (represented by 1) have no tributaries; when two of them join, a second-order stream (represented by 2) is formed. Similarly, when two second-order streams join, a third-order stream (represented by 3) is formed. Finally, two third-order streams join to form a fourth-order stream (represented by 4). *(Illustration by Jeff Dixon.)*

order, and so on. The watersheds of those tributaries are called, respectively, first-order, second-order, and third-order watersheds (see Figure 2.6).

Living in a River

Life, through more than 3 billion years of evolution on Earth, has found ways to adapt to the most challenging environments: utter darkness at the bottom of the sea, high temperature water, extreme acidity, and extreme aridity. No wonder, then, that plants and animals have been able to adapt to life in running water environments. Some adaptations that make existence in the stream environment possible are described below.

Adaptation to Flow

Flow is the defining condition of life in running waters. Currents exert a force (drag) in the direction of flow on anything exposed to the moving water. To remain stationary, to hold one's position in the face of moving water, is challenging. Anyone who has ever tried to walk across a fast-moving stream can attest to that. From the point of view of an organism, whether a plant, a fish, or an invertebrate, to keep from being swept downstream requires the expenditure of energy, or some other strategy.

One characteristic of flowing water that stream dwellers use to their advantage is a layer of reduced water velocity near the channel sides and bottom. Here is how it works: when a moving fluid and a stationary solid surface are in contact, the layer of water molecules closest to the solid "stick" to it, and therefore have a velocity of zero. This zero-velocity layer drags on adjacent layers, and the result is a velocity profile that looks like a "J," with the lowest velocities closest to the bottom and the greatest near the top of the water column.

Thus, a layer of reduced velocity, and therefore reduced drag, extends some distance out from the substrate. Body shapes of benthic organisms—some of which spend much of their lives attached to stationary objects in moving water—are often flattened or streamlined to take advantage of this refuge from the force of moving water. Examples include the stonefly larva and the water penny (see Figure 2.7).

Other aquatic macroinvertebrates demonstrate a broad array of other adaptations to flow. The larvae of certain caddisfly species (order Trichoptera) construct cases made of sand particles and pieces of organic debris, which are anchored to the substrate with silk. In addition to giving the animal a refuge from predation,

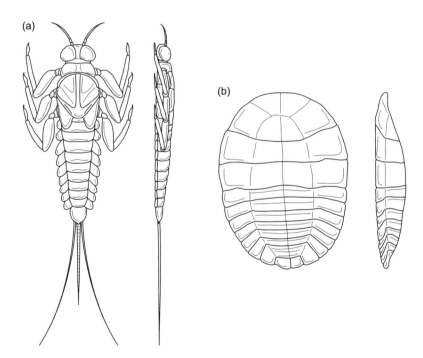

Figure 2.7 Adaptation to flow conditions. Mayfly (left), water penny beetle (right). *(Illustration by Jeff Dixon. Mayfly adapted from Hynes 1979.)*

they help keep it from being dislodged by the current. Other species, for example, stonefly nymphs (Plecoptera), have claws to help them hang on to the rocks they occupy. Still others (black flies and water pennies, for example) have suction-cup-like attachment disks.

Behavioral adaptations to increased flow include moving to refuges (protected areas) within the stream during or just before floods (fishes); leaving the stream altogether when rainfall begins (giant waterbugs); and moving away from exposed surfaces during floods (caddisfly larvae).

River plants also have developed adaptations to flow. Most aquatic macrophytes (plants large enough to be seen easily without a microscope or magnifying lens) have large root systems for both anchoring and rapid resprouting after floods. Willows and cottonwoods, tree species that often grow in the flood channel, have stems that can withstand bending and narrow leaves to reduce drag during floods. Adaptations notwithstanding, few macrophytes can persist in high-velocity areas of the streambed.

Adaptations to Flow Variability

River flows can vary by four to five orders of magnitude, from raging floods to a mere trickle. Sometimes these variations are predictable, for example, in climates with distinct wet and dry seasons. In other regions, their timing is unpredictable, even though the number of years on average between, say, large floods can be calculated with a reasonable degree of confidence. The organisms that live in rivers have evolved a number of strategies for avoiding the risks and taking advantage of the opportunities presented by living in such a dynamic environment.

In some river systems with predictable floods, for example, in the spring, many biological events are synchronized with those floods. Some salmonids (a family of fish that includes salmon and trout) time their spawning or the hatching of fry to avoid the season most likely to have floods. Some amphipods do so as well. Some riparian willows and cottonwoods release their seeds just as the spring floods begin to recede. In river systems with random flooding, some fish species spawn just after floods to ensure that their fry have adequate water.

A drift-flight cycle appears to be a means by which some invertebrates deal with flows and recover from floods. Aquatic insect larvae may "enter the drift"—meaning they let go of the substrate and let the current take them—accidentally as the result of a flood, or purposefully to find "greener pastures," or to escape predation. Whatever the reason, a downstream movement of larvae is quite apparent. Upstream populations persist over time because some insects, for example, the *Baetis* mayflies, in their adult, winged stage tend to fly upstream, thus regaining their position.

For macrophytes that inhabit backwaters and river margins, floods and droughts mean alternating high and low water levels. Plants may either not be rooted or be loosely rooted to be able to float off in higher water. Floating mats of macrophytes are seen in many river systems, as well as wetlands and lakes. By floating on top of the water in areas not exposed to currents, the problem of being submerged by high water disappears. Another adaptation is the development of special highly porous tissues called aerenchyma; these elongated tissues allow movement of atmospheric oxygen to submerged roots. Other metabolic and morphological adaptations are seen in macrophytes as well (these are discussed at greater length in Chapter 3).

River Biota

The same general groups of plants, animals, and other organisms (see Chapter 1) inhabit rivers as wetlands and lakes. Relatively few species are uniquely riverine compared with the large number that inhabit freshwater systems generally. This fact attests to the interconnectedness of freshwater aquatic habitats.

Trophic Relationships

In all natural ecosystems, whether terrestrial or aquatic, some organisms provide the food energy that is consumed by both themselves and all other organisms. These are known variously as producers or autotrophs (literally, self-feeders); for the most part, these organisms are green plants. In the process of photosynthesis, they combine atmospheric carbon dioxide (which may be dissolved in water) and water to produce a variety of carbohydrates, liberating oxygen as a by-product. Thus, they use solar energy to create potential energy, which is stored in chemical bonds and then converted to mechanical energy and heat in the process of respiration. The energy that autotrophs do not use for their own growth and reproduction is called net primary production, and it is the food basis for all heterotrophs (organisms that do not produce their own food energy through photosynthesis; literally, "other feeders").

Autotrophs in river systems. River autotrophs can be classified into three major groups according to their form and physical habitat: periphyton, phytoplankton, and macrophytes (see Chapter 1). The periphyton and phytoplankton together account for the majority of autotrophic production in most rivers. Where their concentrations are high, for example, in the Pantanal of Brazil (see Chapter 3) during certain seasons, they can exert significant influence on levels of dissolved oxygen and carbon dioxide through their photosynthesis, respiration, and eventual decomposition.

The abundance of periphyton and phytoplankton is limited in rivers by a number of factors. A common limiting factor is light. In headwater streams in forested ecoregions, shade from trees can reduce light input into a stream and significantly limit algal populations. In lowland streams, where shade from trees is not a factor, turbidity—whether from sediment or phytoplankton—may limit algal production to a relatively shallow depth or surface photic zone. Another limiting factor for algal production commonly encountered in freshwater systems, including rivers, is the chemical element phosphorus, a major nutrient required by plants. Nitrogen is less commonly limiting in freshwater systems. Finally, periphyton abundance can be reduced by the physical wear and tear associated with floods.

Sources of energy. The autotrophic algae described above, along with some photosynthesizing bacteria, are the sources of food energy produced in the stream. Such energy is termed autochthonous energy. Other energy, allochthonous energy, comes from terrestrial plants, as well as from the terrestrial food webs that they support. Like all green plants, terrestrial producers use solar energy to build carbohydrate molecules that store the sun's energy, which is transformed into proteins, fats, and other molecules in the plant tissues. Stored energy is extracted for use by the plants through the process of respiration. When leaves fall, when insects die, when bears do what they do in the woods, when pollen blows, when tree branches blow off—some of these materials either fall into, are blown into, or are washed into streams. Even the organic (carbon-based) material in soil is lost to streams. As rains fall and percolate through the soils, soluble organic compounds move with them through the near-surface groundwater and into the stream system.

Allochthonous energy becomes available to fish and other higher organisms in the river's community of life through the activity of the decomposer community. Stream ecologists recognize three major forms of organic material (potential food) that enters the aquatic food web from allochthonous sources: coarse particulate organic material (CPOM), fine particulate organic material (FPOM), and dissolved organic material (DOM).

CPOM consists of particles of organic material from about 1 mm diameter up to the size of a tree trunk. Typical CPOM sources include the leaves and needles, fruits, and branches of trees; dead aquatic macrophytes; feces of larger animals; pollen and seeds; and carcasses. CPOM typically is not consumed as such, but rather is broken down through physical degradation (abrasion, dissolution) as well as the activity of decomposers and detritivores. Many invertebrates specialize in different activities related to the breaking down of CPOM and have mouthparts designed for shredding, scraping, gouging, or mining. Typically, the invertebrates do not begin their work until the CPOM has been "softened up" by bacteria and fungi, which colonize it and break down some of the tougher materials over a time frame of hours to days. They also add nutritional value to it for the invertebrate detritivores.

Much of the FPOM in a river is the product of physical degradation of CPOM, as well as the activities of the invertebrate processors. These invertebrate detritivores not only break off small pieces during processing, but their fecal matter also becomes a source of FPOM. FPOM itself is food for a wide variety of organisms, from the microscopic or near-microscopic to the relatively large. Numerous small animals have evolved strategies for finding and capturing FPOM. Many aquatic insects deploy something like a fine net into the current, with which they collect drifting FPOM. In some cases (for example, the blackfly larvae), the net is actually a body part; in others, it is spun from silk (for example, some caddisfly species). Mussels and freshwater clams consume FPOM and phytoplankton by filtering water internally. Some annelid worms as well as some insect larvae burrow in FPOM-rich fine sediment deposits. Fish, birds, and mammals consume these detritivores. The allochthonous energy sources thus supply food to higher trophic levels.

Allochthonous energy sources also supply food to the microbial community. Bacteria utilize DOM from groundwater sources and from the dissolving of easily soluble compounds in CPOM and FPOM. Protozoans and other small consumers then feed on the bacteria. The food derived from DOM may continue to cycle in the microbial food web or, to the extent that such consumers are themselves consumed by larger organisms, it may pass to higher trophic levels and larger organisms.

The proportion of the food energy budget coming from terrestrial (allochthonous) versus in-stream (autochthonous) sources varies from headwaters to mouth along the length of a river. This is one of the main points of the River Continuum Concept (see below). But climatic differences also play a role. Shading, leaf litter inputs, rates of decomposition of organic material, and sunlight intensity are all variables that change with climate and that critically affect the relative contributions of instream versus land-based production.

Decomposers. Decomposers, strictly speaking, are those organisms that consume dead organic material, breaking it down into an inorganic form. Decomposers in rivers include primarily fungi and bacteria. Together with the detritivores, they are an important link between allochthonous sources of energy and higher trophic levels in the aquatic food web. The detritivores include micro- and macroinvertebrates, as well as a few species of fish and other vertebrates.

Heterotrophs. Heterotrophs account for all the other trophic levels in the aquatic food web. Heterotrophs that feed on autotrophs or producers are known as primary consumers; heterotrophs that prey on other heterotrophs are known as secondary, tertiary, or higher-level consumers. In aquatic systems, some organisms fit neatly into such a scheme, occupying only one trophic level, while others might feed at several levels. Some macroinvertebrates, amphibians, and fishes move from one trophic level to another at different life stages. Certain crayfish, for example, are herbivores while small and then move toward a more predatory style as they grow older and larger.

For ecologists, it is useful to group heterotrophs into guilds, or functional feeding groups, to help understand their role in the food web. Both macroinvertebrates and fish have been thus classified. Many of these organisms have evolved either specialized body parts or behaviors that help scientists determine their functional feeding group. Many invertebrate shredders, for example, have mouth parts designed for tearing, cutting, and ripping. Invertebrate predators like the mighty hellgrammite may have large pincers for grasping and immobilizing their prey.

. . . .

Macroinvertebrate Feeding Groups

Macroinvertebrates can be categorized by four feeding groups: grazers, shredders, collectors, and predators.

Grazers. Organisms that feed on a thin layer of organic life, primarily periphyton but including microscopic animals, bacteria, and detritus, that coats underwater surfaces. They typically have mouth parts that enable them to shear, grind, or rasp off attached organic material from submerged surfaces. Examples include water pennies and snails. Water pennies are a good example of an organism adapted to life in running water, as their bodies are flattened and streamlined, allowing them to resist being dislodged by swift currents.

Shredders. Organisms that feed upon CPOM, primarily leaves. They prefer leaves that have been in the water long enough to have been colonized by fungi and bacteria. Examples include crane flies (Diptera), some stonefly species (Plecoptera), and scuds (order Amphipoda, family Gammaridae). Some species of true flies (Diptera), beetles (Coleoptera), and caddisflies (Trichoptera) are known as shredder-gougers; these organisms feed on woody CPOM.

Collectors. Organisms that feed upon FPOM. Collectors can be further distinguished on the basis of whether they collect FPOM that is suspended in the water (filterer-collector) or that is in sediments that have settled to the bottom. The hydropsychid caddisflies (order Trichoptera, family Hydropsychidae) spin nets that they use to capture FPOM as it drifts by. Sediment dwellers, such as oligochaetes worms (see Chapter 1), burrow through sediments and ingest them, digesting the organic material and passing the rest through their digestive tract.

Predators. Organisms that feed upon animal prey, primarily other invertebrates but occasionally small amphibians or fishes. Predators typically have mouthparts designed for piercing and biting. Examples include many dragonfly (Odonata) larvae as well as numerous species of dobsonflies (Megaloptera), stoneflies (Plecoptera), caddisflies (Trichoptera), and beetles (Coleoptera).

. . . .

Collectors either spin nets or extend net-like body parts into the current to collect fine particulate matter. Grazers, such as some snail species, have mouthparts designed for scraping and rasping. Two caveats should be kept in mind: first, some species do not fit neatly into one particular group, but rather may feed opportunistically, eating whatever is abundant at the moment. Second, some species change groups at different life stages. These caveats apply to fish as well as to invertebrates.

Fishes, like the invertebrates, have evolved feeding specializations that are reflected in both behavioral repertoires and body parts, for example, jaws, teeth, and digestive system. Indeed, body shape itself has evolved according to feeding preferences in many species. Predators whose specialty is lying in wait and darting out to seize their prey—pike and gar, for example—have highly streamlined, torpedo-like bodies. Some bottom feeders—suckers, for example—have mouths that are oriented downward like vacuum cleaner hoses. Fish that feed on insects floating on the surface have upward-tilting mouths. Body shape in some cases, however, corresponds more to the general habitat preferences—for example, pool versus riffle, top versus bottom of water column—of the species in question, rather than to feeding preferences. The following fish feeding groups are found in North and South American rivers (after Allan 1995):

- Piscivore
- Benthic invertebrate feeder
- Surface and water column feeder
- Generalized invertebrate feeder
- Planktivore
- Herbivore-detritivore
- Omnivore
- Parasite

River Ecosystems

River scientists have labored to understand river ecosystems in a way that would allow them to discover patterns that would apply to all rivers. Several of the leading models are described in this section. Each scientific model focuses on a different aspect of rivers and sheds light on different factors influencing the life of the river. The River Continuum Concept, for example, elucidates the influence of physical factors that change as one moves longitudinally down a river, from headwaters to lowlands. The Flood Pulse Concept broadens our understanding to include the lateral influence of floodplains on river life. A vertical dimension is brought in by including the influence of the hyporheic zone. Studies relating the river system to landscape processes within the watershed broaden our understanding of how processes on different scales can interact. Change over time is emphasized in still other studies, so that the observed patterns of life in a river are related to the specific history of that river.

Each river is unique, and its community of life represents a unique response to its particular conditions. Scientists traditionally have been concerned with describing and understanding pristine river systems, but virtually none are left in the world. The study of river ecology is like trying to draw a moving object. River scientists cannot get rivers to "sit still" because larger processes such as climate change, bioinvasions, and societal manipulations of rivers continue. The lowland sections in particular have been extensively modified in most large rivers, making higher-order streams difficult to study. One consequence is a great many studies of small streams, but not many of large rivers. This is related to ease of sampling as well.

Community persistence and stability. Stream communities in undisturbed, unpolluted conditions seem to be able to persist quite nicely in the face of considerable environmental variability. Floods and droughts occur regularly, if not predictably, over medium to long timescales, and stream organisms are adapted to survive and recover from such disturbances. Stream communities thus are said to have a high degree of resilience: even though a flood may drastically reduce populations of benthic organisms and fish, they can reproduce and recolonize relatively quickly. The connectedness of river systems over large areas makes it possible for aquatic organisms to move from one tributary to another relatively easily.

If a disturbance is severe, or if new habitat appears (for example, a new meltwater stream from a glacier), successional processes occur. As with succession in terrestrial habitats, succession is initiated by pioneer species, which are then replaced, at least partially, by other species that outcompete them. In general, species with short lifespans are early colonizers. Flying insects have an advantage over invertebrates that are exclusively aquatic, if there are geographic barriers, and will therefore be the first to recolonize a highly disturbed stream. Among the phytoplankton and periphyton, diatoms seem to be early colonizers. Macrophytes colonizing a newly created bar or island after a flood also demonstrate a pattern of succession, with hardwoods appearing last.

The persistence and (therefore) predictability of river biotic communities over time leads to the question of why species composition differs in one part of a river versus another, upstream versus downstream. This question is addressed by the River Continuum Concept.

The River Continuum Concept. The River Continuum Concept (RCC) relates river biota and ecosystem processes to stream order, and thus to those aspects of the stream that vary predictably with stream order (see Figure 2.8). At the same

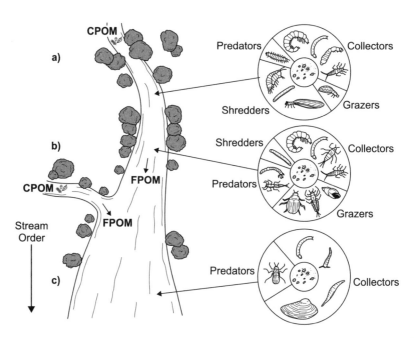

Figure 2.8 The River Continuum Concept relates position in river network, food source, and composition of biotic community. *(Illustration by Jeff Dixon. Adapted from Trayler 2000.)*

time, the RCC—as the name suggests—recognizes that stream systems are a continuum, with headwater streams (orders one through three) grading into medium rivers (orders four through six), which grade into large lowland rivers (seventh order and above), as one moves downstream. The biological community in the stream is adapted to the prevailing conditions at any point on the stream. It is a community in an ever-changing balance, or dynamic equilibrium, with the physical conditions of the stream, which are also in dynamic equilibrium.

Headwater streams have the following characteristics relative to downstream reaches: steeper slope, lower discharge, larger sediment particle size (gravel, cobble, boulders), shallower water, narrower channel, more shade or less direct sunlight, greater inputs of terrestrial material such as leaves, and lower water temperatures. Therefore in the headwaters, there should be less instream production—less food energy produced in the stream by periphyton, phytoplankton, and macrophytes. This is because of heavy shading during the growing season (at least in watersheds dominated by deciduous forest) and the lack of suitable conditions (substrate, currents) for growth of macrophytes. Thus, the stream community uses more food energy than it produces. At the same time, there are relatively large inputs of material from terrestrial sources. These make up the deficit.

In mid-size streams, the stream is wider, so relatively little is shaded. The water is still relatively shallow and turbidity is low, so instream production can dominate: the stream autotrophs produce more food than they use. Because less of the stream is under a tree canopy, leaf inputs are reduced, and the terrestrial energy sources become less important. However, there is considerable importation of FPOM from upstream sources. These sources can include CPOM that has broken down, both mechanically and through the action of shredders, as well as feces from upstream consumers. With deposition of fine sediments in areas of reduced current (along margins of the stream), conditions become suitable for growth of macrophytes, and these add their contribution to instream energy production. The stream community produces more food than it can consume, at least seasonally.

In large rivers, even less shading occurs, but production by periphyton and phytoplankton falls off because of increased turbidity (which limits light penetration) and greater depths. The stream becomes dependent again on instream energy and food energy from upstream, mostly in the form of FPOM. The reduced habitat diversity characteristic of large rivers means species diversity is also reduced relative to mid-size streams.

The composition of the invertebrate community and to a lesser degree the fish community reflects the changing conditions. In the headwaters, the functional feeding group, shredders, makes up a relatively large proportion of invertebrates. The close connection between river environment and riparian zone in the headwaters ensures that plenty of FPOM enters the stream, and FPOM is produced by the shredders in breaking down CPOM as well; therefore, the RCC predicts a relatively large proportion of invertebrates as collectors. The relatively few grazers reflect the general lack of periphyton.

In the mid-size rivers, the RCC predicts that the increased algal production will increase the proportion of invertebrate grazers; shredders will correspondingly decline. The mid-size rivers have a relatively high species diversity both of invertebrates and fish, because they act as an ecotone between the headwaters and the large lowland rivers. The diversity of food supports a diversity of invertebrates, which in turn supports a diversity of fish.

In large rivers, a major food source is FPOM from upstream. The many deposits of fine sediments are full of organic materials. Both these conditions favor collectors. Although turbidity and depth may limit algal production, higher concentrations of nutrients might result in seasonal population increases, for example, in low water conditions.

The proportion of invertebrates as predators is not predicted to change much as one moves down the river. The fish community shifts from cold-water species to warm-water species, and fishes that need coarse substrates are seldom found in large lowland rivers.

The RCC has stimulated a great deal of research and discussion. It is clear that not all river systems follow the model as described—for example, high-latitude rivers, rivers in extremely arid regions, and rivers that have been subject to human modification. Nonetheless, the concept has advanced our understanding of how conditions in and along the river affect the life within its banks.

The Flood-Pulse Concept. While the RCC focuses exclusively on a continuum of life that changes along the river in tandem with changing hydrologic and morphological conditions, the Flood-Pulse Concept (FPC) calls attention to the importance of the regular inundation of the floodplain to the life in the river. Floodplains vary in significance among river systems, but broad, well-defined floodplains are most common in the lowland sections, where rivers are largest. Some river systems have enormous floodplains; the lower Mississippi River had about 38,000 mi^2 (almost 100,000 km^2) before large-scale leveeing and channelization.

The FPC calls attention to the ecological linkages between the river and its floodplain, drawing upon studies of large tropical rivers with lengthy, predictable periods of inundation. Floodplains in such rivers resemble enormous wetlands and frequently are studied as such. Much of the aquatic habitat created by the floods is lentic (stillwater, or lake-like) as opposed to lotic. The Pantanal in South America (see Chapter 3) is an example of such a system.

At the beginning of the high water period, which in some rivers may last for half the year, a littoral (shoreline or beach) zone gradually moves upward in elevation, inundating terrestrial vegetation and making its nutrients and organic material available to aquatic species. Fish and aquatic invertebrates move into the expanding aquatic habitat. The reproductive cycles of many fishes in such rivers are timed to correspond to this cyclical flood event. Thus, the food webs of floodplains and their associated habitats (wetlands, uplands, and oxbow lakes) are deeply intertwined with those of the main river channel. The FPC complements the RCC by including this lateral expansion of aquatic habitat and its contributions of nutrients and organic material.

Patch dynamics. Another concept that complements the RCC is that of patch dynamics, which considers river habitats as mosaics of small patches of distinct physical environments. In the pool-riffle sequence in alluvial streams, pools are viewed as one patch and riffles as a distinctly different patch. To some degree, the occurrence of patches is random. A given biotic community at any particular point along a river would reflect the local patch characteristics and would be only partly predictable from the RCC. Patch dynamics emphasizes the importance of scale: the RCC is a large-scale model, but the particulars of local habitat patches are small-scale phenomena.

Nutrient spiraling. The Nutrient Spiraling Concept (NSC) takes a familiar ecosystem concept—nutrient cycling—and gives it a form appropriate to running water habitats. In terrestrial and lentic (lake) aquatic habitats, ecologists have identified material cycles for the major nutrients required by plants and animals: nitrogen, phosphorus, silicon, and other chemicals. These nutrients in an inorganic (mineral) form are absorbed in the soil by plant roots, incorporated into plant biomass (organic), and then returned to the soil (broken down back into mineral form) by decomposers when plants die. In lotic systems, this nutrient cycling takes place in running water, so the cycling involves some downstream transport of nutrients before the cycle is completed. The path traveled by a nutrient atom in passing through the cycle can therefore be visualized as a spiral. The rate at which nutrients are removed from the system depends on both flow rates and cycling rates. Ultimately nutrients carried by rivers are discharged into the marine environment.

Rivers and Their Terrestrial Context

River systems are embedded in terrestrial ecoregions, and the terrestrial biota and ecological processes influence the biodiversity of the river systems within them. Aquatic ecoregions have been proposed as a way to classify geographic areas in terms of expected communities of aquatic plants and animals. Aquatic ecoregions may be contiguous with watershed boundaries in some cases, but diverge in others. Ecoregional setting is one of several influences on the biodiversity of streams and rivers, along with land use history. No universally agreed-upon classification of regions for freshwaters exists.

The watershed has been proposed as an ecologically defined geographic region and is sometimes conflated with the ecoregion concept. With respect to freshwater aquatic biomes, the interconnected system of freshwaters within a major watershed is probably the closest analog to a terrestrial ecoregion. It is not the watershed per se that makes up the aquatic environment, as a watershed is a terrestrial region, but the river system draining the watershed that creates this environment.

For example, the snail darter (see Plate III), a small benthic fish made famous by the Tellico Dam controversy in the 1970s on the Little Tennessee River, was thought to have been condemned to extinction by construction of the dam. However, populations were later found elsewhere in the larger Tennessee River basin. Also in the Tennessee River basin are a number of endemic mussels that are not found in adjacent river systems, illustrating the fact that for these organisms at least the logic of the river system as ecoregion makes sense.

The connectedness of a river system means that aquatic species found in one part of a watershed are likely to be found elsewhere in the watershed. Species that are strictly aquatic, like fishes (as opposed to, say, insects that are aquatic in early life stages but then emerge from the water and fly), may evolve in a single river system, dispersing to adjacent river systems only very slowly or not at all. The distribution of freshwater fishes, therefore, reflects better than almost any other group of organisms the distributional patterns reflected in the biogeographic realms originally delineated according to the occurrence of terrestrial birds and mammals. Some freshwater scientists subdivide these six realms to define freshwater regions within them, but here it will suffice to list the six realms and note their dominant fish families.

- The Nearctic Region includes North America south to the southern end of the Mexican plateau. Major river systems include the Mississippi, the world's third largest, as well as the MacKenzie, the Yukon, and the Columbia. The freshwater biota of this region is more thoroughly studied and better described than any other. Fourteen families of primary freshwater fishes are known in this region and almost 1,000 species. The dominant families are the minnows and carps (Cyprinidae), suckers (Catostomidae), North American freshwater catfishes (Ictaluridae), perches (Percidae), and sunfishes (Centrarchidae).
- The Neotropical Region consists of Central America southward to include the whole of South America. This region contains the largest river system in the world, the Amazon, as well as the Rio de la Plata, the Orinoco, and the São Francisco. With 32 families of primary freshwater fishes that include more than 2,500 species, this region has the richest freshwater fish fauna of any of the biogeographical realms. Dominant taxa include about 1,300 species belonging to 13 families of catfishes (order Siluriform) and more than 1,200 species belonging to eight families of the characins (order Characiformes: tetras, paranhas, pencilfishes, hatchetfishes, and headstanders). Other taxa represented in this realm include order Gymnotiformes (electric fishes), Cichlidae (angelfishes, oscars, discus fishes), and numerous members of marine taxa that have invaded and become acclimated to freshwaters. There are no representatives of the minnow family (Cyprinidae) or sucker family (Catostomidae) in this realm.
- The Palearctic Region is a huge region composed of Europe, much of central and northern Asia including northern China, the Middle East, and northern Africa. Major river systems are the Ob-Irtysch, Yenisei, Lena, Chang Jiang (Yangtze), Amur, and Volga, as well as the Nile. Nonetheless, it has relatively few primary freshwater fish families. The dominant families include loaches (Cobiditidae) and the minnow family (Cyprinidae).

- The Ethiopian Region includes sub-Saharan Africa and Madagascar as well as adjacent portions of the Arabian Peninsula. This region hosts the world's second-largest river system, the Congo, as well as several other major river systems, including the Niger, the Zambesi, and the Orange. This is a species-rich region with 27 families of primary freshwater fishes and about 2,000 species, including primary and secondary fishes. Taxa represented include the minnows (Cyprinidae), characins, and catfishes (Siluriforms). A number of archaic primary freshwater fish species occur as well. These living fossils include the Polyptiridae (bichirs) and members of the genus Protopterus (African lungfishes).
- The Oriental or Indo-Malay Region includes the Indian subcontinent, southern China, Southeast Asia, the Philippines, and the East Indies. Major river systems include the Ganges, the Indus, the Brahmaputra, the Mekong, and the Irrawaddy. The region contains 28 families of primary freshwater fishes, including 12 families of catfishes (order Siluriforms), cyprinids, loaches (Cobitidae), and snakeheads (Channidae).
- The Australian Region includes Australia, New Zealand, Tasmania, and New Guinea. This region, which includes only one major river system, the Murray-Darling in Australia, has only three species of primary freshwater fishes: a Ceratodontid (the Australian lungfish), and two species of the family Osteoglossidae (Arowanas). The rest of the several hundred species in this region are not primary freshwater fishes.

While the distribution of fish taxa reflects well the delineation of biogeographic realms, most other inhabitants of the river biome are more cosmopolitan—that is, taxonomic groups are spread throughout the world. With reference to the structure and functioning of lotic communities, differences between rivers in one biogeographic realm and those in another may be less important than headwaters versus lowlands or one ecoregional setting versus another along the same river. Accordingly, it is unlikely to be fruitful to examine in depth a major river from each biogeographic realm. However, there are broad differences between rivers at different latitudes. The following sections profile a tropical river (the Amazon), a high-latitude river (the Amur), and a pair of mid-latitude rivers, the New and the Tennessee. What stands out in comparing these river systems is perhaps the uniqueness of each despite the fact that they share many features common to all rivers.

The Amazon River

The Amazon is a giant among rivers. It sends more water in a year to the sea than the six next biggest rivers combined: the Congo, the Yangstze, the Orinoco, the Brahmaputra, the Yenisei, and the Rio de la Plata. Every one of these is an enormous river, yet the Amazon dwarfs them all. It is a river of superlatives: it has more species of fishes than any other (an estimated 3,000); it drains the world's largest intact tropical forest; it has the largest watershed; and when it floods, it floods the largest area. It discharges so much fresh water and sediment into the Atlantic that the color and salinity of the ocean is altered for 200 mi (321 km) out to sea. It is also a river undergoing rapid ecological change and facing an uncertain future.

Characteristics of the River and Its Watershed

The watershed of the Amazon has been estimated to be between 2.4 and 2.9 million mi^2 (6.1 million and 7.5 million km^2). It includes much of Brazil, and parts of Paraguay, Bolivia, Peru, Ecuador, Colombia, and Venezuela. The shape of the watershed is unusual in that once the headwater rivers leave the Andes, the water drops very little on its trip to the sea. In other words, except for the Andes, much of the basin and all of the central basin is quite low (about 1,000 ft/300 m or less), giving the mainstem Amazon an extremely low slope—less than an inch per mile. One reason is that the Amazon is a river that has reversed course. It once flowed east to west, and discharged into the Pacific. Then, when the geologically young Andes arose, the river was blocked to the west and had nowhere to go but east.

It is difficult to generalize about the characteristics of the Amazon basin (see Figure 2.9). Not only is it immense, but it contains a tremendous variety of landscapes, terrestrial ecoregions, and landforms, from the Andes rising to 22,000 ft (6,700 m) in the west, to millions of acres of wetlands scarcely above sea level. Land cover is as follows: forest, 73 percent; grassland, savanna, and shrubland, 10 percent; wetlands, 8 percent; cropland, 14 percent; dryland area, 6 percent;

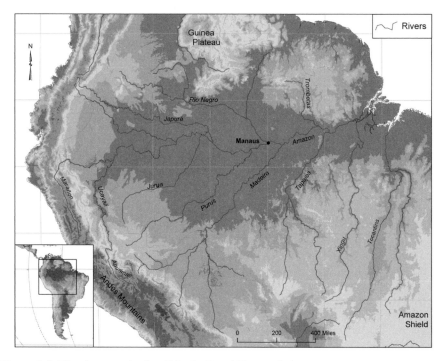

Figure 2.9 The Amazon basin. *(Map by Bernd Kuennecke.)*

and urban/industrial area, less than 1 percent. The vast majority of the basin is dense, relatively unbroken tropical rainforest.

The river has its source in the glaciers of the Peruvian Andes, 3,900 river mi (6,375 river km) from the sea. That source, a thin sheet of clear water flowing across a rock face on the slopes of Nevado Mismi, an 18,362 ft (5,597 m) mountain not far from Lake Titicaca, was only recently pinpointed by using a global positioning system (GPS). The source was once considered to be a headwater of the Rio Marañon, but more recently the Rio Apurimac has been given that honor. The Apurimac is a tributary of the Ucayali, which joins with the Marañon to form the Amazon. The mainstem river, which was formed by the Ucayali and the Marañon, was formerly known (and still is locally) as the Solimões as far down as Manaus, where the Rio Negro joins from the north. Only thereafter was it considered the Amazon.

Downstream from the confluence of the Ucayali and the Marañon, the Amazon is joined by the Jurua and Purus from the south and the Japura from the north. Then the Rio Negro joins from the north, and several hundred miles downstream, another major tributary joins from the south: the Madeira. Continuing toward the sea, the Tombetas, the Tapajos, the Xingu, and the Tocantins add their flow, along with countless smaller tributaries. These major tributaries are very large rivers themselves; the Rio Negro has a larger mean average annual discharge—more than 3.5 million ft^3/sec (100,000 m^3/sec)—than any other river in the world except the Amazon.

One reason for the very large discharge of the Amazon is the climate of the Amazon basin. Rainfall varies from one area to another, but the average rainfall is about 79 in (2,000 mm). Some parts of the basin (particularly the eastern slopes of the Andes, where orographic precipitation occurs) receive as much as 26 ft (8,000 mm) per year. Evapotranspiration rates are high, due to the equatorial temperatures and sunlight intensity, but this rainfall still generates an enormous amount of runoff.

Temperatures across the basin vary little, not surprisingly because it straddles the Equator (so there is no latitude effect) and there is little altitude change until the extreme western margins in the Andes. Ascending the slopes, the usual drop in temperature is seen ($-3.5°$ F per 1,000 ft or $-6.49°$ C/1,000 m). Because of the equatorial position, there is little seasonality in temperature, which averages $75°–82°$ F ($24°–28°$ C) year-round. Rainfall in the central part (but only in the central part) of the basin shows a definite seasonality, with a dry season usually from June to November.

By tradition, the tributaries of the Amazon are grouped into three types. The whitewater rivers are so called because they carry large loads of fine sediments and so look cloudy—not really white but more like latte. These rivers drain the actively eroding Andes. The black-water rivers, actually tea-colored, are stained with acids formed by dissolving organic material and have a relatively low pH. They drain the heavily forested valleys and uplands of the Guiana Plateau. This region is geologically old and its soils have long since lost their easily dissolved or eroded minerals. The clear-water rivers, entering the Amazon from the south, are also nutrient poor but lack the DOM that stain the black-water rivers. They drain the Amazon Shield, another geologically old formation now covered with either moist or dry forest, or savanna, called *cerrado*. The differences between black-water rivers and clear-water rivers are primarily related to soil types in the upper parts of their watersheds.

The Amazon and, because of the extremely flat and level watershed, its tributaries too, flood predictably once a year, and the flood is of impressive magnitude. Atmospheric moisture from the Atlantic makes its way west, and beginning in fall, that moisture is released, as air rises up the Andes and comes down as rain. The month of maximum rainfall is different in different parts of the watershed: in the Bolivian Andes, it is January; in the Ecuadoran Andes, it is April; in Manaus, roughly halfway from the Andes to the sea, it is March. The contributions of the various tributaries are not "in sync," but the net result is a slow oscillation of the river level over the course of a year, with low water in the middle sections of the Amazon (around Manaus) occurring in October through January, and high water in April through July. Flood levels are variable from one year to the next, and also on the different river sections; but the timing is dependable enough that many aquatic organisms, from fish to caimans, time spawning and other important life history events to the rising and falling of the water.

Upstream of Manaus in the section that is sometimes called the Solimões, the water level may rise roughly 33–49 ft (10–15 m). This is not to suggest that the Amazon mainstem is at all "flashy" in the sense of rapid and extreme rises and fall in discharge and water level. In fact, changes in the river's discharge are strikingly moderate; high water and low water discharges differ only by about a factor of three. Streamflow variations of four or five orders of magnitude (that is, high flows 1,000 to 10,000 times greater than low flows) are more typical for rivers, particularly smaller ones. The very low slope of the river is responsible for the "piling up" of flood waters in the central basin, leading to the extreme variations in water level.

Such a water level increase in a region that has little topographic relief and in a river whose channel slope is extremely low means that the floodplain forests are inundated for weeks at a time every year. The presence or absence of flooding and length of the period of inundation leads to the local naming of three different forest types in the lower Amazon basin: two that are flooded regularly (the *varzea* and the *igapo*) and the dry land rainforest of *terra firme*. The floodplains along the Amazon and its whitewater tributaries are called *varzea*, although this term is also used in a broad sense to denote any of the forested floodplains subject to regular inundation. These areas are nourished annually by the influx of fine sediments, nutrients, and organic material brought in by the rising muddy waters of rivers draining the Andes and their foothills. In turn, these flooded forests nourish the fish and other living organisms in the river. On black-water and clear-water streams, such as the Rio Negro and others draining the old, weathered soils north and south of the Amazon, the flooded land is called *igapo*. Flooding here gives rise to a different community of organisms, both terrestrial and aquatic, than the *varzea*. In both cases, however, the floodplain is inundated during the high water period, sometimes to impressive depths.

In low water, a number of lakes are left behind by the retreating water. These then undergo ecological transformations as evaporation shrinks them, plants grow and decompose, and oxygen conditions fluctuate. These lakes are emptied of fishes as soon as the river reclaims them in high water season, and the fish head for greener pastures in the flooded forests.

River Biota

The Amazon forest, lying within the Neotropic biogeographic realm, is fabulously rich in biological diversity. The same is true of the river system—not surprisingly, because the river is interwoven with the forest to such an extent that at times it is not clear where the dividing line is. By one estimate, 30 percent of water discharged to the sea by the Amazon has been through the floodplain. Another reason for the high degree of aquatic biodiversity is the variety of habitats, from mountainous whitewater rivers, black-water rivers, and clear-water rivers, to the normal diversity of habitats described in the RCC. To attempt anything like a comprehensive description of all the life of the Amazon River would take several large volumes. The description here is limited to a few general observations and focuses on a few groups of plants, invertebrates, fishes, and other vertebrates.

Plants. For the photosynthetic organisms, the different river habitats (including floodplain lakes, as they are seasonally part of the river) provide a variety of conditions and challenges. This variety results in tremendous diversity of plants. In general, four types of vegetation are found in the rivers of Amazonia: algae (phytoplankton and periphyton), aquatic herbaceous plants, terrestrial herbaceous plants, and woody plants (the floodplain forest). The mix of plant species appearing at any particular place and time on the river, including its floodplain and floodplain lakes, is determined by gradients of inundation (for how long, how deep, how often) as well as human influences. These gradients manifest—subtly, in this flat region—as zones of elevation, substrate stability (determined by erosion and sedimentation), and succession. In addition, the many different habitats of the Amazon basin give rise to different plant associations.

The main river channel in general is not a welcoming place for plants, in any of the three river types. While species diversity is high, phytoplankton biomass in the rivers and streams of the Amazon is relatively low despite warm temperatures and intense sunlight. In black-water and clear-water streams, production is relatively low because of a lack of nutrients. In low-order forest streams with full tree canopy coverage, light limits phytoplankton production. In whitewater streams, which do have ample nutrients, turbidity limits light penetration and thus phytoplankton production.

In the whitewater rivers, including the mainstem Amazon, floating mats (also called floating meadows) of vegetation are common and harbor whole communities of plants and animals. Grasses (water paspalum and aleman grass, primarily) spread quickly in low water on exposed sediments in the floodplains. When high water returns, these colonies develop floating stems with hanging root masses and essentially "lift off" with the rising water. They are joined by other floating plants like eared watermoss, which like a number of Amazon plants has become an aquatic nuisance species when introduced elsewhere. The floating mats are colonized extensively by insects and their predators, and their underside provides habitat for extensive growth of the periphyton, which then supports a complex web of zooplankton and fishes. Carnivorous bladderworts on the floating mats produce underwater bladders that implode on contact with microcrustaceans or insect larvae, engulfing and digesting the victims.

Channel bars are rapidly changing deposition and erosional features. Soils, such as they are, are sandy. They may be subject to inundation for most of the year, and dry out only in the lowest part of the low water period. Here, annual terrestrial plants grow and spread rapidly when the opportunity arises. Some, as described above, can become free-floating and form part of the plant matrix of floating mats when the water rises. Steeper river banks are subject to strong currents and wave action. Mexican crowngrass, aleman grass, and water paspalum are among the relatively few species found on such banks. Fewer still are found on unstable steep banks subject to sloughing.

Floodplain lake beds and low-lying flats are not subject to strong currents or erosion except by waves, and receive only very fine deposits during their period of inundation, which is lengthy. During their brief (one to two month) dry period, they support a relatively low diversity of plants, including grasses like West Indian marsh grass (a major aquatic nuisance species, particularly in sugar-cane growing areas where it has been introduced), and sedges including burrhead sedge (a nuisance species in areas of North America where it has been introduced). As the water rises, more aquatic species, including some species of wild rice (genus *Oryza*), as well as free-floating plants, succeed the primarily terrestrial species. Floodplain lakes during the low water period and floodplain backwater areas during flood can support large populations of phytoplankton. Some lakes become eutrophic, as huge populations of diatoms and cyanobacteria take advantage of reduced turbidity and lack of turbulence.

Low-lying swales are similar to the low-lying flats except that they do not dry out as completely, retaining enough soil moisture to support some aquatic free-floating plants even during the low water period.

The floodplains, despite lengthy annual inundation, do support shrubs and trees. The floodplain forests show a heterogeneity that reflects the complex and dynamic landforms that they inhabit. The floodplain lakes and abandoned channel swales are too wet most of the time to support woody vegetation, but natural levees (where the river initially deposits its greatest sediment load as it enters the floodplain) and transitional areas to the *terra firme* are dry long enough to do so. The dynamic nature of the landscape created by the rivers as they meander back and forth ensures that forests in a number of different stages of succession can be found in proximity.

The Central Amazon varzea tends to be predominantly evergreen moist tropical forest with a mosaic of plant communities whose composition is determined by seral stage, soil characteristics, and frequency and length of inundation. Early seral shrubs include Lythraceae, a member of the loosestrife family; Iporuru; and a willow, *Salix martiana*. This willow, like many in the flooded forest, develops adventitious roots as the water rises to transport oxygen from the air to submerged parts of the plant. Later, successional woody plants include early successional trees whose seeds are dispersed by fishes; a spiny palm whose seeds can withstand being submerged for 300 days; and a tree, *Cecropia latiloba*, which is effective at colonizing the nutrient-rich floodplains of the whitewater rivers by virtue of being highly flood- and submergence-tolerant as well as tolerant of intense sunlight and sediment deposition. It has submergence-tolerant seeds and can grow vertically quickly. Late successional trees include species from the genera *Ceiba* or *Chorisia* (beautifully flowering trees that include the kapok), *Eschweilera*, *Hura* (the juice of one species is used to poison darts), *Spondias*, and *Virola*, a member of the nutmeg family. Igapo forests, inundated by nutrient-poor waters, have fewer species than the varzea forests inundated by the whitewater streams. In both types of flooded forests, many life-cycle events of trees are controlled by the timing of the flood pulse, including fruiting, flowering, and seed production, dispersal, and germination.

Invertebrates. Invertebrate communities have been studied at sites in the Andean headwaters of the Amazon. Given that the substrate is relatively coarse gravel and cobble, it is not surprising that mayflies, stoneflies, and caddisflies were the most abundant species, along with some beetles. The first three of these groups prefer fast-moving, cold, well-oxygenated waters characteristic of mountain streams and rivers.

The benthic invertebrates of the lowland rivers have not been much studied, apparently on the assumption that the shifting sand substrate is not suitable for most bottom-dwelling species. This sandy substrate extends upstream 2,485 river mi (4,000 river km) from the mouth of the Amazon. The low-order headwater streams of the black- and clear-water tributaries, however, support a variety of insects. Midge flies (Diptera, family Chironomidae) are the most diverse insects in these streams, along with black flies (Diptera) and caddisflies (Trichoptera). Cold-water species, for example, stoneflies (Plecoptera), are largely absent, as are molluscs, which are apparently limited by the low levels of dissolved minerals. An important food chain in these streams is fungi feeding on allochthonous material from the surrounding forests, being consumed by small chironomids, which are then consumed by a variety of predators.

While conditions for benthic invertebrates are unsuitable in the large lowland rivers, the floodplain lakes provide conditions sometimes favorable to relatively high densities of invertebrates (up to almost 1,200 benthic animals per ft^2). However, when the floods recede, many of these lakes develop low oxygen levels, particularly at the bottom, and this reduces invertebrate numbers.

By some accounts, the greatest diversity of aquatic invertebrates occurs on and around the floating mats found in the flooded *varzea* and in the large rivers. The floating macrophytes in them support periphyton among their roots, and this provides the basis for a thriving invertebrate community sometimes referred to as the perizoon. The perizoon also colonizes submerged or emergent macrophytes during the inundation of the floodplains. Organisms such as water mites, copepods and ostracods, water fleas, and fly larvae are present in great numbers, particularly in mats in whitewater streams. They provide food for many species of fish and larger invertebrate predators. The mats also harbor snails of the genus *Biomphalaria*, which carries the parasitic blood fluke that causes schistosomiasis, a disease that afflicts many in the tropics.

Several floodplain invertebrates are important in the river food web due to their large numbers. Freshwater shrimp are particularly numerous in black- and clearwater streams. The Amazon River prawn, however, inhabits whitewater streams, and its abundance makes it important prey for many fishes. Apple snails are large and numerous and are an important part of the diet not only of several fish species but also of caimans and some birds. The snail-kite has a beak that is shaped to fit inside the shells of the snails, which it consumes almost exclusively. Another important invertebrate link in the whitewater river and floodplain food web, by virtue of its great numbers, is a species of mayfly. In its larval stage, it is efficient at tunneling into dead wood, hastening its decomposition.

Fishes. The number of species of fishes in the Amazon basin is variously estimated at between 2,000 and 4,000. In general, the freshwater fishes of South America are "remarkable on the one hand for [their] small number of ancestral stocks and on the other for richness and endemism in certain groups" (Darlington 1982, p. 69). Two such groups are the characids (order Cypriniformes), which make up 40 percent of the species in the basin, and catfishes (order Siluriformes), which account for 25 percent. In addition, 11 other orders of fishes are present, representing 21 families, some of which contain many species. As with the other lifeforms in the Amazon basin, the diversity of fishes can be attributed to the tremendous diversity of habitats in the region. Unexpectedly, the species diversity of the black-water rivers differs little from that of the whitewater rivers. In both systems, species diversity is higher than would be found in temperate climate streams of the same order.

Physical factors that account for the distribution of fishes include water chemistry, oxygen levels, and presence or absence of lotic conditions. Some black-water river species are relatively intolerant of the higher dissolved mineral levels in the whitewater streams. This is thought to lead to a higher degree of endemism within some tributaries, as the whitewater mainstem Amazon would present an insurmountable barrier to dispersal into other black-water streams for such species. Many species without such a limitation, however, use the mainstem Amazon as a dispersal pathway, either actively migrating or using the currents to disperse their young.

Some fishes avoid the mainstem Amazon and its large, lotic tributaries altogether, dispersing into the *varzea* or *igapo* at high water and remaining in floodplain lakes (*lagos*) during low water. Such species are likely to have evolved either morphological or behavioral means of dealing with low-oxygen conditions, which are common in the *lagos* during the low water season. The Black Acara, for example, can survive low-oxygen conditions by breathing air into its stomach. Other taxa absorb oxygen by breathing air into their mouths, gill cavity, swim bladder, and intestines, which have special adaptations to make this effective. Yet others, particularly very small fish, inhabit a shallow surface zone in the *lagos* that is oxygenated through contact with air. Others still move toward open water and away from macrophyte stands at night, when periphytic production ceases to oxygenate the water; during the day, they move back to benefit from the oxygen provided by the periphyton living on the macrophytes, and by the macrophytes themselves, many of which have physiological adaptations to move atmospheric oxygen into their submerged parts.

The characids are small- to medium-size fishes that have a variety of habits and habitats and probably reach their maximum speciosity in the Amazon basin. There are 15 genera of characids in the Amazon basin; the group includes many species familiar to tropical fish fanciers, such as the tetras. It also contains probably the most famous of the Amazon's fishes, the piranhas (family Serrasalmidae), of which 28 species occur in Amazonia.

These aggressive fishes, 6–10 in long (15–25 cm), have well-developed jaws and extraordinarily sharp teeth as befits a carnivore. However, recent research has shown that they may eat fruit and other vegetable matter at some life stages. Nonetheless, it is as carnivores that they affect the other fishes in the Amazon. Their impact is substantial because of their great numbers, which also make them important in the diets of dolphins, birds, and humans in Amazonia. Their feeding habits, including massed attacks, are particularly effective during the crowded conditions of low water in the *lagos*.

The piranhas have cousins among the Serrasalmidae that eat a widely varied diet, including seeds and nuts, fruit, insects and insect larvae, and other fish and invertebrates. The tambaqui is the second-largest Amazonian scaled fish, and has teeth "like a horse," allowing it to crush seeds and nuts in the flooded forests. This is a popular game fish among the local inhabitants of Amazonia, whose fishing pressure, unfortunately, appears to be causing its numbers to decline.

The catfishes (order Siluriformes), although distributed widely around the world, reach their greatest diversity of form and habits in the Amazon. Catfishes come very large and very small, narrow and wide, conspicuously striped and camouflaged as pieces of woody debris; they are predators, herbivores, detritivores, and omnivores. The largest is locally known as *piraiba*. Individuals can grow to about 10 ft (3 m) and weigh more than 330 lbs (150 kg), and while they are primarily piscivores, stomach contents have included parts of monkeys. By some reports, they prey on humans at times.

At the other extreme in the catfish family is the *candirú* or toothpick fish. Also known as the vampire fish, it is a parasite that lives on the blood of other fishes, which it gets by swimming up into their gills and biting an artery. This fish reaches a maximum of 2 in (5 cm). It has been known to swim up any human orifice it can find and is particularly attracted to urine; once lodged in the human urethra, it is nearly impossible to remove except surgically. For this reason it is feared by Amazonians. Swimming nude in the Amazon is not recommended.

Besides the piranhas and the toothpick fish, another denizen of the Amazon to avoid incidental contact with is the electric eel, a member of the order Gymnotiformes. The use of electric fields to locate prey is characteristic of the gymnotiforms, but only the electric eel among freshwater fish has the additional capability of generating and discharging high-voltage shocks. Sufficient to stun or kill the eel's prey, they can temporarily disable and perhaps kill a grown man. The gymnotiforms of the Amazon also include more than 50 kinds of knife fishes, many of which are associated with the floating meadows and varzea lakes, as they tolerate low-oxygen waters.

The cichlids are another group of fishes that has burst into a fireworks display of diversity in the Amazon, with well over 100 species in this family of the order Perciformes. Some are favorites of aquarium owners, such as the Oscar fish and the Angel fishes.

Although the other groups are not so specious as the characids, the siluriforms, and the cichlids, some stand out by virtue of playing a major ecological role. Herrings and anchovies (order Clupeiformes) are mostly marine, but the 13 species in the Amazon are extremely numerous and feed on zooplankters. The Aripaima (order Osteoglossiformes) is the largest scaled fish in the Amazon and is a top predator in the ecosystem. It can grow to more than 9 ft (3 m) in length.

Other vertebrates. While the ecological roles of fishes in the Amazon are relatively well known, the aquatic food webs of Amazonia include a number of other important vertebrates about whose roles less is known: caimans, freshwater dolphins, manatees, otters, and turtles. The numbers of many of these organisms have been reduced drastically by hunting. The largest caiman, the black caiman, has been almost entirely eliminated from the system—hunted for its skins—with unknown but presumably far-reaching ecological impacts. The black caiman is a fearsome predator that can reach well over 13 ft (4 m) in length and 3 ft (1 m) in width.

The world's largest freshwater dolphin, the Amazon River dolphin, known as the boto to Brazilians, lives in the Amazon and its tributaries. This mammal is sometimes called the pink dolphin (some but not all adults are quite pink). It lives near the bottoms of rivers, where it consumes fish, turtles, and crabs. Another dolphin, the gray dolphin or tucuxi, shares the waters of the basin. Similar to the marine bottlenose dolphin in shape, but slightly smaller, this mammal consumes a wide range of fishes, at least 28 species belonging to 11 families. Both dolphins are protected by national legislation in Brazil, but both are considered threatened by habitat loss and pollution.

Another exclusively aquatic mammal in the Amazon River and its tributaries is the Amazon manatee, which can weigh up to 1,000 lbs (about 450 kg). Like all manatees, this species is completely herbivorous. It eats aquatic plants in the flood-plain lakes and backwaters and during the dry season lives off stored fat while inhabiting the main river channels. Manatee numbers have declined due to hunting.

In keeping with the Amazon's reputation as the river of superlatives, the giant otter (its resident otter) is the largest of the world's otters. This carnivorous, mostly aquatic member of the weasel family inhabits a variety of slow-moving clear-water habitats. It can grow to nearly 6 ft (2 m) in length and about 65 lb (30 kg) in weight on a diet of fish, primarily characids. The smaller Amazonian river otter inhabits lower-order streams and backwaters of the floodplain and has a varied diet that includes, in addition to fishes, crustaceans such as the freshwater shrimp, amphib-ians, reptiles, birds, and small mammals. With no natural predators, both otters are threatened by hunting (for pelts) and habitat loss. The World Conservation Union (IUCN) considers the Amazonian River otter to be the most endangered otter spe-cies in the world, a dubious superlative indeed.

Numerous species of turtles in Amazonia are largely aquatic. The largest among them is the giant Amazonian River turtle, whose carapace typically reaches 25–30 in (64–80 cm) in length. The best-known, perhaps, is the unique-looking *mat-amata*, which occupies an ecological niche similar to that of the alligator snapper of North America. Both are opportunistic piscivores that use camouflage and lie in wait for their prey on the muddy bottom of mostly lentic waters. Their strategies for capturing fish differ slightly, however, with the matamata striking out with its jaws open while sucking in water, in effect vacuuming the fish into its mouth.

Many other reptiles and amphibians occupy primarily the floodplains of the Amazon system, but few if any live exclusively in the water. Probably the most fa-mous reptile in the system is the anaconda, which, while large, is probably not as big as some reports would make it. Its maximum verified length is about 40 ft (12 m)—by no means small—but less than the 65 ft (20 m) sometimes cited. It fre-quents the rivers because its prey is mostly aquatic. It lives on fish, turtles, caimans, and capybaras, which are large rodents that live in the floodplains.

Problems and Prospect

The Amazon River and its forested watersheds for many years seemed easily to repel man's puny attempts to settle and exploit them. Since the European coloniza-tion of the Neotropics, however, the human impact has been growing steadily more disruptive. Prior to European settlement, many parts of the forest supported large populations of native peoples, but contact with European culture caused high death rates caused by enslavement, forced labor, relocation, disease, and misery. Even today, many of the less accessible parts of Amazonia support a much reduced human population. Nonetheless, with modern technologies and ever-increasing demand for new materials and land, human society is altering the system, reducing biodiversity in the process.

The greatest threats to the Amazon forest come from logging, land-clearing for agriculture, pollution from mining, and climate change. What affects broad land areas in the watershed must also affect the aquatic life it harbors. Forest-clearing, whether for agriculture, urbanization, or logging, causes habitat loss and chokes streams with sediment. The Brazilian Amazon region has received as immigrants from other regions of Brazil more than 1 million farm households since the 1970s. The number of head of cattle in the Amazon region has doubled since the early 1990s. This influx of immigrants, human and hoofed, has come in response to different pressures. The promise of free land, government relocation schemes, government dam- and road-building, and hopes of quick riches by gold mining or mining of biological resources all have contributed. New roads in particular have accelerated environmental decline by providing unprecedented access to the forest interior.

The net result has been land-clearing on a massive scale, and with it habitat loss, erosion, and sedimentation. Widespread, small-scale, and often illegal gold mining has result in mercury contamination of aquatic food chains. Human settlements along the rivers have created sewage pollution. Overfishing and hunting have led to population declines particularly in the larger fishes, as well as some mammals and reptiles.

Concern continues to rise over infrastructure projects planned by the Brazilian government. Several additional highways and a series of large dams on the Xingu River are in the works, although these developments are contested by indigenous groups and environmental campaigners. Already several large dams are operating in the Tocantins River watershed, and scores more are under construction or planned. Two high dams are planned on the Madeira River. Dams are disruptive of physical and ecological processes in rivers, in rough proportion to their size.

The other chief concern regarding the Amazon region is the potential (and perhaps present) impact of climate change. In 2007, the region was in its second year of unusually intense drought. Some scientists believe that the drought may reflect climate changes under way worldwide. Because the forest generates so much of its own precipitation, the concern is that a prolonged drought might set off a vicious cycle of intensifying drought, wildfires, and desertification, potentially leading to collapse of the present ecosystem, which is intertwined with the river and its inhabitants, as this chapter has demonstrated.

The Amur River

The Amur River basin lies between roughly 45° and 55° N latitude and 120° and 140° E longitude in eastern Asia, entirely within the Palearctic biogeographic realm. Its 745,000 mi^2 (1,929,955 km^2) watershed is the 10th or 11th largest in the

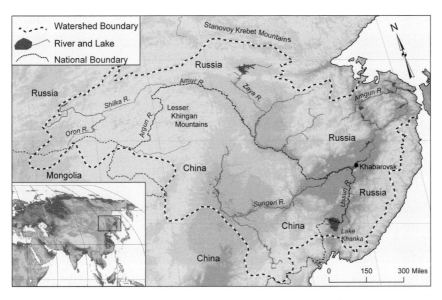

Figure 2.10 The Amur River and its basin. *(Map by Bernd Kuennecke.)*

world, and its average annual discharge of 441,000 ft^3/sec (12,500 m^3/sec) at the mouth is the 14th largest. The length of the Amur proper—from its mouth at the Sea of Okhotsk to the confluence of the Shilka and the Argun—is 1,770 mi (2850 km), but from there to the source of the Argun is half again as far (another 900 mi/ 1,500 km) (see Figure 2.10). It is the largest free-flowing river remaining in the world—for the moment: China and Russia reportedly are planning a series of hydroelectric dams on the mainstem Amur. The Amur watershed lies mostly in northern China and far eastern Russia, and includes northern Manchuria and parts of Mongolia. Much of the basin is relatively low in elevation (less than 500 ft or approximately 150 m), although it is fringed with mountains, including the Stanovoy Khrebet range to the north.

Characteristics of the River and Its Watershed

Today, about 54 percent of the basin is forested (this represents about two-thirds of the original forest cover); 9 percent is grassland, savanna, and shrubland; 4.4 percent is wetlands; 18.4 percent is cropland; and 2.6 percent is urban and industrial land. About one-third is dryland (deserts and desert-like lands). Satellite images reveal that much of the cultivated land is in the floodplain and undoubtedly has replaced some wetlands.

The river system is traditionally divided into three parts (starting at the mouth and going upstream): the Nizhniy Amur, the Sredniy Amur, and the Verkhniv Amur. The Nizhniy Amur runs from the mouth to approximately the city of Khabarovsk, just downstream from the confluence with the Ussuri, an important tributary that drains Lake Khanka. In the Nizhniy Amur section, relief is low and the river channel divides

into a complex and changing system of braided channels and floodplain lakes and wetlands hydrologically connected to the river and frequently inundated during the warm,
high water months. Such a channel form is typical of rivers with regular, extreme fluctuations in discharge. The channel bed material is primarily sand and silt. The river is
joined by one large tributary, the Amgun, and several smaller ones, in this section.

Farther upstream, the Sredniy Amur extends to where the Zeya River joins
from the north. At this point, the Russian city of Blagoveshchensk is on the north
bank, and the Chinese city of Heihe is on the south. The river channel through this
section is also braided and has a well-developed floodplain, except in a 95 mi (153 km)
stretch where the river has cut a narrow gorge through the Lesser Khingan
Mountains. The substrate is sand and silt except where the river passes through
the Burein Mountains, where a stony substrate and bedrock predominate. A major
tributary, the Sungari River, enters from the south in this stretch. The watershed
of this large river (almost 1,250 mi/2,000 km long) accounts for more than one-
fourth of the Amur's drainage area.

The Verkhniy Amur extends upstream from the confluence with the Zeya River
to the beginning of the Amur, at the confluence of the Shilka and the Argun. Ghen-
gis Khan is supposed to have been born near the banks of the Oron River, one of
the two major tributaries of the Shilka in Mongolia. Channel morphology and substrate are typical of rivers in more mountainous regions, with single channels, relatively steeper gradient, and larger sediment particle size. In and upstream of this
section, the Amur and its tributaries wander across the high, moderately dissected
plateaus of Mongolia and Southern Siberia. In narrow valleys with steep slopes
covered with boreal forest, the Amur and its upper tributaries, the Shika, Igodo,
Onon, Khoylar, and Argun, have carved passages and formed floodplains replete
with meander scars.

Climate in the basin is characterized generally as humid continental or subarctic. The subarctic climate regions have severe, dry winters and cool summers. The
humid continental areas are characterized by severe, dry winters but humid, warm
summers. The winter months are dominated by a cold, dry Siberian high-pressure
system, so there is little snow. One consequence is that, without a heavy insulating
blanket of snow, the low temperatures freeze the soil, the lakes, and the rivers over
much of the watershed. Many fishes that migrate up the Amur to spawn and feed
in its floodplain lakes must overwinter in the mainstem river itself, as the lakes tend
to freeze to the bottom. Much of the basin is underlain by permafrost.

Another consequence of the dry winters is that snowmelt, while not insignificant (15–20 percent of runoff), contributes considerably less to the annual runoff of
the Amur than does rainfall. The tributaries and lakes are covered with ice from
November through April or May. The cool to moderate summers receive 10–20 in
(25–50 cm) of rainfall, which comes from Pacific moisture, often associated with
typhoons.

The climatic pattern results in enormous seasonal fluctuations in flow on the
Amur. Near Khabarovsk, winter discharge may be as low as 7,000 ft^3/sec (about

200 m^3/sec) or less, since the tributaries are frozen. In summer flood, discharge may be as high as 1.4 million ft^3/sec (about 40,000 m^3/sec). These annual extreme low and high flows, together with freezing conditions and ice floes, make for channel instability and a relatively high level of disturbance for the lotic ecosystem. Fish populations vary greatly from one year to the next because of the seasonal fluctuations in water levels and the extreme climate conditions that freeze lakes and tributaries in winter.

The Biota of the Amur River

Plants. In the headwaters, and indeed throughout the Amur, relatively few aquatic or semiaquatic macrophytes occur, because of the great annual fluctuations in water level. While phytoplankters are scarce in the headwaters, a vigorous growth of the periphyton is possible in the warm season during which time the tree canopy is not completely closed. In the middle river, significant instream primary production is by algae and mosses. In the lower reaches, because of the combination of increased turbidity, depth, and turbulence, as well as lack of suitable substrate, the periphyton community is much reduced. However, sometimes quite high concentrations of phytoplankton are found in the lower river (as measured by concentration of chlorophyll *a*) in the summer, either during low water or when water levels are declining so that floodplain lakes drain into the river. The majority of the phytoplankters are diatoms and green algae, accompanied in summer by large numbers of cyanobacteria. It is thought that the lakes are the source of the high concentrations of phytoplankton, as many of them become so eutrophic that massive fish kills are common. Interestingly, during the period in which the river is frozen over (about five months for the mainstem), a community of cryophilic (ice-loving) microorganisms forms on the bottom of the ice. The diatom *Aulacosira islandica*, which not only survives but thrives on the sharply reduced light under thick ice, is the dominant photosynthesizing species.

Invertebrates. The Amur is reported to be outstanding among Russian rivers not only for its fish diversity but also for its invertebrate diversity. As expected from the RCC, the Amur's headwater tributaries receive considerable organic material from terrestrial sources. Substrate particle size is relatively large (gravel and pebbles) compared with downstream reaches. As the RCC suggests, these streams have a relatively high proportion of benthic invertebrates as shredders. Represented orders include true flies (Diptera), mayflies (Ephemeroptera), stoneflies (Plecoptera), and caddisflies (Trichoptera). Dipterans include the aquatic larvae of Blephariceridae, Chironomidae, and Simuliidae. The Blepharicerids are net-winged midges that live in mountain streams, with their steep slopes and rocky substrate. They are equipped with "suctorial disks" or what essentially are suction cups on their undersides, which help keep them anchored in swift currents. The Chironomids are midge fly larvae, and the Simuliids—the biting black flies—are cursed by outdoor workers and adventurers throughout the far north. In the forested headwaters, two mussel species occur in large colonies.

In the middle reaches of the river, changes in the benthic community consistent with a shift toward autochthonous production is observed, with fewer shredders and more collectors and grazers among the invertebrates. In the floodplain sections of the lower river, benthic organisms include molluscs—including some rare species of mussels; a diverse assemblage of insect larvae from the orders Diptera, Odanata, Ephemeroptera, Plecoptera, and Trichoptera; and aquatic worms. In low water conditions in the summer, algal blooms may result in a shift in the insect larva assemblage as ephemeropteran, plecopteran, and trichopteran numbers decline while populations of chironomids increase. In general, the composition of the benthic community reflects substrate and hydrology.

Fishes. Some 120 species of fish are reported from the Amur River system, of which seven are endemic and three are introduced. This total is significantly higher than any other Northern Eurasian river system. In the upstream reaches and tributaries, the Amur grayling, the taimen, and the lenok are common. The Mongolian taimen is the world's largest salmonid (salmonids include salmon, trout, and whitefish; see Chapter 1), reaching up to about 6 ft (180 cm) in total length and weighing more than 200 lb (about 100 kg). While this is impressive, the taimen has grown in legend: by some accounts, they can reach 30 ft (about 10 m) in length and weigh 4 tons.

In the middle and lower river are two species of sturgeons, including one, the Amur sturgeon, that is listed as endangered on the Red List of Threatened and Endangered Species by the IUCN. This once-plentiful fish can reach about 10 ft (300 cm) total length and 660 lb (190 kg), but overfishing combined with relatively low population resilience (it reproduces but slowly and therefore its population cannot "bounce back" after being depleted) has led to a sharp decline in its numbers.

In floodplain reaches of the river and the floodplain lakes, both upstream and downstream, common species include several carp species, Amur pike, and Amur catfish. In these habitats, several rare species appear, including the black Amur bream, the Chinese perch, and the snakehead.

As one might expect from the fact that the Amur is (so far) undammed in the mainstem, large runs of migratory fishes occur—among them Pacific salmon, lampreys, and a spring run of European smelt.

Problems and Prospect

The human population is growing rapidly in the broad floodplain valleys of the Amur, particularly on the Chinese side, resulting in an intensification of fishing pressure and environmental degradation from agriculture, logging, and mining. As a result, since the 1960s, fish populations have been declining. Logging in the upper tributaries has caused sediment loading, reducing the spawning habitat of migratory salmon. Pollution from urban centers and especially nutrient pollution from agriculture has led, during unusual summer low water conditions, to hypereu-trophication—explosive algae blooms—and the resulting low-oxygen conditions have caused mass mortality of both fishes and invertebrates, particularly in the sys-tem's lakes. In Lake Khanka, for example, biodiversity has recently been reported to be declining rapidly due to eutrophication from agriculture and aquaculture, combined with overfishing.

The decline of certain commercially valuable fish stocks has resulted, at least on the Russian side, in the passage of prohibitions or limitations on catches of stur-geons and salmon. Efforts at habitat restoration, particularly spawning habitat for salmon, are being made. Such efforts aimed at migratory salmon are bound to be in vain, however, if Russia and China follow through on their reported plans to build four hydroelectric dams and reservoirs on the mainstem Amur (that is, not on tributaries). Such modification of the system will predictably decimate the salmon populations and bring about significant change to the aquatic ecosystem.

Rivers of Southern Appalachia: The Tennessee and the New

The southern Appalachians are an old region from a geological standpoint. The mountains, once high and jagged, have been worn down by wind, rain, ice, and the patient activity of billions of trees and other plants over the millennia. The resulting landscape is one of softly rounded mountains, long valleys, sparkling mountain streams, graying barns, and fields with haybales. In the heart of this region are the headwaters of two large rivers that ultimately flow into the ocean via the same route, but by widely divergent paths: the New and the Upper Tennessee (see Figure 2.11).

The valley of the New arises in the high mountains of west-central North Caro-lina and extends northward, following the river's winding path into Virginia, then out again into West Virginia, where it carves a deep gorge. Just before heading into the gorge, the New is joined by its largest tributary, the Greenbrier, which flows southward into the New, draining a long narrow watershed that extends approxi-mately 165 mi (265 km) northward deep into West Virginia. Emerging from the gorge, the New River meets the Gauley and forms the Kanawha, a tributary of the Ohio. The Ohio ultimately heads southwest and is one of the Mississippi's largest tributaries. Water transported by the New from its headwaters in North Carolina ultimately spills into the Gulf of Mexico.

The New is approximately 320 mi (515 km) long; its watershed encompasses almost 7,000 mi^2 (18,130 km^2), and its average daily flow at Kanawha Falls, just upstream from its confluence with the Gauley, is about 12,000 ft^3/sec (340 m^3/sec). The New cuts across the long ridges and valleys of the Appalachians, which tend to run southwest to northeast. This is one piece of evidence among others that have convinced some geologists that the New is in fact very old—maybe the oldest river in North America and second-oldest in the world. While its age is questionable,

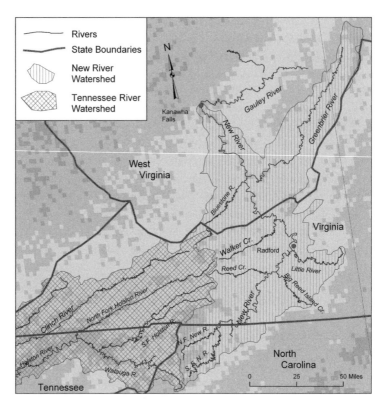

Figure 2.11 Adjacent river systems: the Upper Tennessee and the New. *(Map by Bernd Kuennecke.)*

it certainly contains some world-class whitewater. A relatively steep gradient for a large river and massive blocks of sandstone blocking the current ensure plenty of turbulence as the river continues to cut through the 800 ft (244 m) deep gorge.

Water from the Tennessee's headwaters also ends up in the Gulf of Mexico, but by a more southerly route. The Tennessee, much larger than the New, is a major tributary of the Ohio, which it joins not far from the Ohio's confluence with the Mississippi. The watershed area is 40,300 mi^2 (104,377 km^2) and the average discharge at its confluence with the Ohio is 68,400 ft^3/sec (1,937 m^3/sec).

The mainstem Tennessee begins at Knoxville, Tennessee, where the Holston and the French Broad join their waters. Not far downstream, the mainstem loses its riverine character for most of its journey to the Ohio River. The river has been transformed into a series of reservoirs through the dam-building of the Tennessee Valley Authority (TVA). So thorough has been TVA's program, that the backwaters of one reservoir practically lap up against the next dam upriver. A short length of river goes from the last dam downstream, the Kentucky Dam, to the Ohio. Even above Knoxville, there are several large reservoirs. But the native diversity of the Tennessee lives on, precariously, in the headwaters streams: the Clinch, the Powell, the Holston, the French Broad, the Little Tennessee, the Hiwassee, and others. In the heart of the Southern Appalachians, the headwaters of these two rivers—the Tennessee and the New—are separated only by narrow mountain ridges.

Both the New and Tennessee lie within the Nearctic biogeographic realm and are both in the Mississippi watershed. Precipitation varies with location and topography but generally is higher in the upper Tennessee watershed (48 in or 122 cm per year in Knoxville, Tennessee) than in the upper and middle New (42 in or 107 cm per year). The headwaters of both rivers lie in several physiographic provinces: the Blue Ridge, the Valley and Ridge, and the Cumberland Plateau for the Tennessee, and the Blue Ridge and Valley and Ridge for the New. The New does eventually cut through the Cumberland Plateau, but not in its headwaters.

The two rivers, although sharing a watershed boundary and ultimately part of the Mississippi River basin, lie in different freshwater aquatic ecoregions, distinguished primarily by native fish distributions. The New is part of the Teays-Old Ohio aquatic ecoregion, which includes the Ohio River and its tributaries. The Teays was an ancient river that flowed, like the New, northwesterly into the ancient Mississippi. At some point in its geologic history, the Teays is thought to have flowed into what is now the St. Lawrence River, and it shares some fish species with that system. The freshwaters of this ecoregion collectively harbor an unusually large number of native fishes, mussels, and crayfish, including many endemics. The New, however, has fewer of each group, for reasons that will be discussed below.

The Tennessee combines with the Cumberland River to the north to define the Tennessee-Cumberland aquatic ecoregion. This river system, like the New, is geologically old. This has allowed plenty of time for speciation to occur in a number of taxa. Moreover, the aquatic ecoregion includes a variety of physiographic provinces, creating a large range of habitat types and environmental conditions. On the basis of these two factors it is not surprising that this ecoregion has the greatest freshwater biodiversity in North America and possibly in the world, at least among temperate freshwater ecoregions.

Physical Description of the Upper Tennessee River

In the Blue Ridge physiographic province, the Tennessee's headwater streams flow westward to northwestward through mostly forested lands. This region includes much of western North Carolina as well as parts of South Carolina and Georgia. Its foundation consists of predominantly igneous and metamorphic rock. Elevations are typically 3,000 ft (914 m) above sea level with individual peaks reaching as high as 6,684 ft (2,037 m) at Mount Mitchell. Gradients are steep and substrate is mostly bedrock and boulders. Autochthonous production is low. Streams draining the Blue Ridge portion of the headwaters include the Nolichucky, the Watauga (a tributary of the South Fork Holston), and parts of the French Broad, the Hiwassee, and the Little Tennessee.

The Ridge and Valley province is dominated by long parallel ridges of sedimentary rock running northeast to southwest. The rivers of this region run along the valley floors, which have a relatively gentle gradient. Substrates are primarily alluvial sands, gravels, and cobble. Agricultural land uses dominate the valleys; the ridges are forested. The shallow rivers meander through the valleys and are more productive than the low-order tributaries on the ridges, since they tend to be wide enough to be unshaded. The Powell, Clinch, and Holston are the main Tennessee River headwater streams draining the Ridge and Valley province.

The Powell and the Clinch, however, both include among their headwaters a number of streams draining the Appalachian Plateau, a region of low mountains, steep narrow hollows, and generally rugged terrain. The Guest River, for example, is a Clinch River tributary that descends from the Appalachian Plateau (sometimes called in this area the Cumberland Plateau). This physiographic province was formed when a large block of sedimentary rock was lifted with little tilt or folding; the dendritic stream network on its surface eroded an extraordinarily complex landscape. The area is mostly forested, since there is relatively little land that can be farmed or built upon. Except for the lowest order streams, gradients are generally moderate, unless waterfalls occur. Channel bottoms are primarily sand and bedrock, and channels are characterized by plenty of riffles and runs.

Extraction of natural resources—timber and coal—has left its mark on many stream channels. When the area was first logged in the nineteenth century, it was common practice to float logs downstream by building and then blowing up "splashdams." These were wooden dams that stored up enough water to flush a large number of logs downstream. This process scoured many channels to bedrock, and it is likely that channel morphology (and aquatic life) is still recovering. Coal mining has changed both the physical character and water quality of streams in some parts of the region, usually to the detriment of aquatic life. Smothering of benthic habitat with fine sediments is a widespread problem.

Physical Description of the New River

The mainstem New begins at the confluence of the North Fork New and the South Fork New in Ashe and Watauga counties, North Carolina. Upstream of this point the New has more than 700 mi^2 (1,813 km^2) of watershed with more than 800 mi (1,287 km) of stream channels. Maximum elevation of the watershed, more than 4,800 ft (1,463 m), is reached at Snake Mountain in North Carolina. The steepest sections of the upper New are encountered in the vicinity of Fries, Virginia, where the river makes the transition from the Blue Ridge Plateau to the Valley and Ridge province. Also near Fries, the New encounters the first two, of four, major dams on the mainstem. Two of the four—Claytor Dam near Radford, Virginia, and Bluestone Dam, near Hinton, West Virginia—create large impoundments and have significant effects on the river environment.

Traveling north-northwest through Virginia, the New alternates between alluvial substrate (silt, sand, gravel, cobble) and bedrock. Depth is generally about 6 ft (2 m) or less, and the channel is several hundred ft (100 m) wide. Rapids form where the river crosses layers of more resistant rock, as for example, where it encounters the Tuscarora sandstone at Big Falls, Virginia. Another major rapids results from stubborn sandstone at the aptly named Sandstone Falls in West Virginia. Below Sandstone Falls, the river continues cutting down while the Appalachian Plateau rises above it, creating the famous New River Gorge with its legendary whitewater.

Large tributaries of the New, upstream to downstream, include the Little River coming in from the east, the Bluestone River from the west, and the Greenbrier River from the north. The Greenbrier is noteworthy for the fact that, despite being 165 mi (265 km) long, it is unimpounded.

A feature considered significant from a biogeographic point of view is Kanawha Falls, now incorporated into the hydroelectric project at Hawk's Nest, West Virginia. Together with Sandstone Falls at the head of the Gorge, and the Gorge itself, Kanawha Falls is considered to be part of the answer to a question that has long puzzled zoogeographers: Why is the New relatively depauperate? Which is to ask, why are there so (relatively) few fish species in the New as compared with other rivers in the region?

Biota of the Upper Tennessee

Plants. Common macrophytes of the river and riparian corridor include water willow, red maple, buttonbush, cottonwood, black gum, American sycamore, and black willow.

Invertebrates. While the Tennessee River is known primarily for its diversity of fishes, invertebrates also reach high levels of diversity there. One such benthic invertebrate group is the crayfishes, of which the Tennessee-Cumberland freshwater aquatic ecoregion has 65 species, including 40 endemic. Especially well-represented are crayfishes of the genera Procambarus, Cambarus, and Orconectes. A member of the last group, the rusty crayfish, has been widely introduced to North American river systems, including the upper Tennessee and New, where it is considered a threat to native crayfishes. Eleven crayfish species, including the Obey crayfish, are rare and protected in Tennessee.

More celebrated than the crayfishes is the Tennessee's striking diversity of freshwater mussels. Before human modifications of the river and the watershed began to reduce their numbers, the Tennessee-Cumberland ecoregion included 125 mussel species, with 100 of those native to the Tennessee. Because most of these mussels require riverine conditions—fast-flowing, shallow, well-aerated water—many have been extirpated from most of the mainstem Tennessee. Mussel species richness is outstanding in the Virginia tributaries of the upper Tennessee— the Clinch, Powell, and Holston—since they are among the few remaining unimpounded streams in the river system. Even here, however, this unique fauna is declining under the effect of multiple threats. Some have become extinct, and

Table 2.2 Mussel Species of the Upper Tennessee River Considered Threatened or Endangered by the U.S. Fish and Wildlife Service

Appalachian monkeyface pearlymussel	*Quadula sparsa*
Birdwing pearlymussel	*Conradilla caelata*
Cracking pearlymussel	*Hemistena lata*
Cumberland bean	*Villosa trabalis*
Cumberlandian combshell	*Epioblasma brevidens*
Cumberland monkeyface pearlymussel	*Quadrula intermedia*
Dromedary pearlymussel	*Dromus dromas*
Fanshell	*Cypogenia stegaria*
Fine-rayed pigtoe	*Fusconaia cuneolus*
Green-blossom pearlymussel	*Epioblasma torulosa gubernaculum*
Little-wing pearlymussel	*Pegias fabula*
Oyster mussel	*Epioblasma capsaeformis*
Pink mucket pearlymussel	*Lampsilis abrupta*
Purple bean	*Villosa perpurpurea*
Rough pigtoe	*Pleurobema plenum*
Rough rabbitsfoot	*Quadrula cylindica strigillata*
Shiny pigtoe	*Fusconaia cor*
Tan riffleshell	*Epioblasma wakeri*

Source: Windsor 2000.

many others are listed as threatened, endangered, or species of concern at either the federal or state levels (see Table 2.2).

Other major groups of benthic invertebrates include isopods, arthropods, snails, and insects. Species of mayflies, stoneflies, dragonflies and damselflies, true bugs, caddisflies, true flies, alderflies, fishflies, and dobsonflies, as well as beetles are all well represented.

Vertebrates. The Tennessee hosts an extraordinary diversity of fishes. Two hundred forty-eight species are present, of which 223 are native. Thirty-two are endemic; one, the American Eel, is catadromous. The darters (family Percidae) are very well represented, with 41 species; other specious families include the cyprinids (minnows and carp, Cyprinidae), suckers (Catostomidae), catfish (Ictaluridae), and the sunfish (Centrachidae).

The headwaters support the greatest diversity. In Virginia, the Clinch, the Powell, and the upper parts of the Holston River system support 117 species (98 native, 16 endemic, and 19 introduced), the highest diversity among Virginia's rivers and far greater than the neighboring New River. In these Virginia headwater streams of the Tennessee, distinct differences in fish assemblages and invertebrate communities are apparent. These differences are attributed to several factors. Drainage area (that is, size of watershed upstream of sampling point) is an important control on the diversity and species mix, as is ecoregion and physiographic province. The most distinct assemblages are found in the cool waters of the South Fork Holston's headwaters in the Blue Ridge physiographic province, which are distinguished by the presence of Tennessee shiner, mirror shiner, flat bullhead, river chub, fantail darter, and Swannanoa darter. A distinctive component of the fish fauna of the Cumberland Plateau tributaries is the Tangerine darter, while the rivers of the Ridge and Valley are characterized by striped shiner, telescope shiner, bluntnose minnow, banded sculpin, golden redhorse, greenside darter, redline darter, stoneroller, whitetail shiner, striped shiner, and snubnose darter.

Some of the Tennessee River's more interesting fishes are the paddlefish, the lake sturgeon, and the snail darter. The American paddlefish (see Figure 2.12) is one of only two species in the family Polyodontidae and is one of the largest fishes found in the Tennessee system. It has a paddle-shaped snout and is considered a "primitive" fish essentially unchanged for perhaps 100 million years. A filter feeder that consumes zooplankton, the paddlefish is under considerable pressure throughout its range due to poaching for its caviar-like eggs. The lake sturgeon is a relative of the paddlefish in the order Acipenseriformes, which contains only the paddlefish and sturgeons. Once common throughout the Tennessee River system, the lake sturgeon is being reintroduced into the headwaters region.

The snail darter (see Plate III) is emblematic of the many small, little known, and economically unimportant endemic fish that populate the Tennessee River system. This 3 in (7.62 cm) fish inhabits gravel shoals and riffles in cool, clear streams such as those found in low-order tributaries. The snail darter probably would have continued to live in obscurity, or more likely perished in obscurity, were it not for two events: first, the passage and subsequent implementation of the U.S. Endangered Species Act, and, second, the planned and subsequently built Tellico Dam

Figure 2.12 The American paddlefish. *(Photo by Steven Zigler, USGS, Upper Midwest Environmental Sciences Center, LaCrosse, WI.)*

Reproductive Strategies of Freshwater Mussels

Mussels reproduce sexually, with females bearing eggs and males fertilizing these eggs with sperm released into the water in the vicinity of a female (see Figure 2.13). The females release the fertilized eggs, now developed into a larval form called a glochidia, into the water in large numbers. To survive, the glochidia must attach, parasitically, to the gills of certain fish 'hosts.' The fish hosts carry the glochidia, which cause them little or no harm, until they mature sufficiently to survive on their own, at which point they drop off. The mussel then begins its life in the substrate. In this way, the mussels disperse into new habitat.

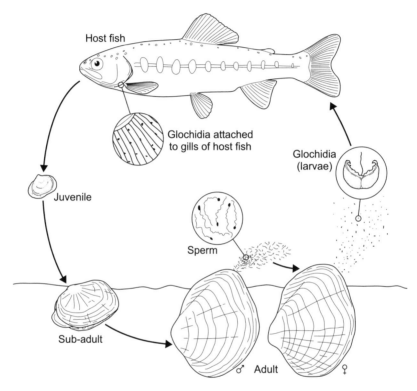

Figure 2.13 The reproductive cycle of the freshwater mussel. *(Illustration by Jeff Dixon.)*

Mussels have developed a number of strategies to get a fish host. Some release the glochidiae into the water where fish will consume them; most are eaten, but some attach to the gills. Other mussels excrete a mixture of mucus and glochidia that appears a bit like a gelatinous worm; when attacked by a fish, it releases glochidia. Some even use a tether on this bait, like a human angler. The most interesting adaptations involve the females' use of mimicry to lure fish close enough to dose them with glochidia. These mussels' mantle tissue, which they can stick out of their shells, looks like prey fish or insect larvae for specific host fishes (see Plate IV). Unfortunately for some of the endangered mussels, if something happens to their host fish species, the mussels are doomed. What has happened to some hosts is that the free-flowing waters and clean coarse substrate many of them require has disappeared, buried under reservoirs created by dams or clogged with fine sediments. Thus, the mussels' best hope of escaping these poor conditions themselves—their fish hosts—are lost to the same conditions.

on the Little Tennessee River. The fish was discovered in streams that would be submerged under the waters of Tellico Reservoir if the dam were built; however, it was listed under the Endangered Species Act as an endangered species in 1975. Supporters of the Tellico Dam were aghast that their project might be derailed to protect a tiny, insignificant (in their view) fish. Through various political and bureaucratic maneuverings, including modification of the regulations under the Endangered Species Act, the dam's supporters eventually won permission to proceed with completion of the dam despite the endangered status of the snail darter. Subsequently, other populations of the snail darter were discovered and still others were established by a stocking program. The controversy involving this humble fish was the focus of considerable media and public attention during the 1970s, however, and made the snail darter one of America's most famous fishes.

The outstandingly diverse fishes and mussels of the Tennessee River systems are interrelated. These two groups of animals coevolved within the same river system, and it is fascinating and delightful to see how their lives are intertwined.

Mussels are filter feeders that live in fast, shallow, free-flowing rivers with clean sand, gravel, or cobble substrate. They are long-lived organisms. In many parts of the Tennessee River system, living specimens and populations are uniformly old. With no young, eventually, without human assistance, they will die out. Recruitment of new individuals to the existing populations is zero, because for many of the mussels, fishes are the key to their reproductive success and continued existence, and in many cases the fishes are gone (see the sidebar Reproductive Strategies of Freshwater Mussels).

Like most aquatic ecoregions of the eastern United States, the Tennessee has a rich diversity of aquatic amphibians; reptiles, including turtles and snakes; birds; and such mammals as beaver, river otter, and muskrat. One group for which the Tennessee basin is particularly noted is salamanders. There are several species that dwell exclusively in streams and rivers, including the shovelnose salamander, whose distribution ranges up and down the eastern edge of the Tennessee River watershed. A member of the family Plethodontidae (the lungless salamanders), this small amphibian lives in rocky highland streams.

Another obligate stream-dweller is from the family Cryptobranchidae and may be the best-known salamander in the Southern Appalachians—the legendary hellbender. A nocturnal hunter that preys primarily on crayfish, the hellbender typically spends the daylight hours hidden under a large rock in the middle of a stream or even a large river. The hellbender, locally known as "grampus," "mud cat," and "Alleghany alligator" among other names, can reach 29 in (74 cm) in length.

Biota of the New

In general, the native terrestrial biota of the New watershed is not markedly different from that of the upper Tennessee River basin, particularly within the same physiographic province. However, for fishes and other freshwater aquatic organisms that do not have a winged or terrestrial life stage, there are significant differences.

The New, as mentioned earlier, is "depauperate" in terms of native fish species—in other words, there are fewer species (not fewer individuals) than one would expect for a watershed and river its size. A mere 46 native species, including eight endemics, inhabit its waters, plus (historically) the catadromous American eel. There is only one centrarchid, the green sunfish. Fisheries managers, anglers, and chance have boosted the number of fish species in the New by another 42, an extraordinarily large proportion of nonnative, introduced species. Introduced or likely introduced are 11 species of cyprinids (minnows and carp); 10 species of centrarchids (sunfishes)—including the highly prized gamefishes largemouth and smallmouth bass; the singular bowfin; a number of prey species such as the alewife apparently introduced to support the introduced predatory game fishes; the golden redhorse, which is native to the Tennessee River tributaries of Virginia; several catfish; striped and white bass of the family Moronidae; several members of the family Percidae, including the walleye; and the ubiquitous rainbow trout and brown trout. Endemics include the bigmouth chub and two other cyprinids, the Bluestone sculpin, the Appalachia darter, the Kanawha darter, and the colorful candy darter (see Plate III).

Compared with the adjacent rivers of the Tennessee drainage, not to mention the remainder of the Kanawha watershed downstream, the New also has a pittance of mussel species (see Table 2.3). The Kanawha (downstream of Kanawha Falls and the New River Gorge) itself has only about 40 native mussel species, and a mere 11 have been found in the New. Most of those species found above Kanawha Falls are rare, populations having declined over recent decades.

It appears that the paucity of native species in the New is the result of a combination of factors. First, downstream barriers prevent upstream movement of fish and other organisms (such as mussel glochidia hitching a ride on a fish) from the greater Mississippi River system: Kanawha Falls, Sandstone Falls, and to some extent the New River Gorge. The view of the New as geographically isolated is supported by the relatively high percentage (17 percent) of endemic fish species.

Table 2.3 Mussel Species of the New River

Elktoe	*Alasmidonta marginata*
Giant floater	*Pyganodon grandis*
Green floater	*Lasmigona subviridis*
Mucket	*Actinonaias ligamentina*
Paper pondshell	*Utterbackia imbecillis*
Pistolgrip	*Tritogonia verrucosa*
Pocketbook	*Lampsilis ovata*
Purple wartyback	*Cyclonaias tuberculata*
Spike	*Elliptio dilatata*
Tennessee heelsplitter	*Lasmigona holstonia*
Wavy-rayed lampmussel	*Lampsilis fasciola*

Source: Appalachian Power Company 2006.

Second, climate may be a factor. Compared with the Tennessee, or indeed to any of the watersheds surrounding the New, the elevation of the New and its watershed is considerably higher, and therefore colder. It has been proposed that during repeated glaciations, the valley of the New became cold enough to extirpate fishes (and perhaps mussels) that were not cold-tolerant. These same species, surviving in the lower elevations of the ancient Teays–Old Ohio system, could not recolonize the New when temperatures became more suitable because of the physical barriers. Moreover, the New's superior elevation meant that it was more likely to lose species to stream capture than to gain them. The upper Tennessee and the New may have traded stream captures in their headwaters, thus commingling species, but in the New, the species may have been extirpated and then not have been able to recolonize.

Prospects for the Two Rivers

The Upper Tennessee. There is an encouragingly high level of attention to conserving the outstanding aquatic diversity of the Tennessee, particularly its fish and mussel species, but one cannot be optimistic about the long-term trends. True, multiple state and federal agencies as well as nongovernmental organizations like the Nature Conservancy are involved in protecting and restoring habitat, breeding fish and mussels for reintroduction, and mitigating pollution. However, in the context of the scope of the problems confronting the Tennessee and its tributaries, these very limited efforts are not likely to increase in the foreseeable future. In the meantime, pollution continues: agriculture, construction, and mining go on, and the degradation goes on, even accelerates, punctuated from time to time by catastrophic accidents. On the North Fork Holston, toxic pollution from the now-defunct Olin Mathieson Corporation chemical plant has been chronic and, together with occasional catastrophic releases of mercury-contaminated sludges into the river, has probably been responsible for the extirpation of several fish species from the Holston, to say nothing of invertebrates. The Clinch, in addition to the usual litany of pollution sources, has been the site of several large chemical spills that had catastrophic impacts on the river for miles downstream. In 1967, the dike of a waste-settling impoundment containing fly ash from a coal-burning power plant broke, releasing highly alkaline sludge into the river. Almost everything in the river for about 12 mi (20 km) downstream was killed. A few years later, the recovery of that stretch of river was set back by a sulfuric acid spill from the same power plant.

In 1998, a truck carrying a toxic chemical overturned and released more than 1,000 gal (3,785 L) of its deadly cargo into the Clinch. The spill turned the river white and killed most aquatic life for about 7 mi (11 km) downstream. Included in the toll were almost 20,000 freshwater mussels, including individuals of three species listed as endangered under the Endangered Species Act. Coal mining continues to affect many of the tributaries of the Clinch, Powell, and Holston.

A more insidious threat to the mussels of the Tennessee is the exotic, invasive zebra mussel (*Dreissena polymorpha*), a native of lakes in southeastern Russia. Introduced into the Great Lakes via ships discharging ballast water into the St. Lawrence river, the zebra mussel has spread through major river systems in the United States and is now in the Tennessee system. Biologists fear that its inevitable spread throughout the river system will bring about the demise of many of the mussel species they are working to conserve.

The New. It is likely that the fauna in the New will continue to change, as the effects of human activity continue and intensify. The New has seen one of the most dramatic changes in fish fauna of any large river system, and changes in species abundance and distribution are accelerating, not stabilizing. The ultimate effects on native species of the large number of introduced fish species remain to be seen. Sections of the river are currently listed as impaired because of the presence of the toxic industrial chemical PCBs (polychlorinated biphenyls), mercury, and several other toxic chemicals. Sedimentation from forestry, agriculture, and construction is a long-term problem, destroying benthic habitat for a variety of species. Indeed, most rivers of the Southern Appalachians are still adjusting morphologically to the huge sediment inputs from over a century ago, when the original forests were first cut. While it seems unlikely that any large new impoundments will be constructed, flow modification by the existing large dams has unknown (because it is unstudied) ramifications. With the additional threats of acid precipitation and climate change, a unique river system's biological integrity, already heavily affected, is unfortunately likely to continue to diminish.

Human Impacts on the River Biome

According to the National Research Council, "Aquatic ecosystems worldwide are being severely altered or destroyed at a rate greater than that at any other time in human history and far faster than they are being restored." It is difficult to imagine that any other biome has been subject to such widespread change and, frankly, degradation as rivers and streams. Many of the alterations seen as improvements by their sponsors—dredging to accommodate shipping, construction of hydroelectric dams, construction of flood protection levees, introduction of game fish species— degrade the river or stream as an aquatic habitat for the organisms living there. They may create benefits for some people, but with the exception of restoration projects, of which there are few, it is all on the negative side of the ledger ecologically. This is to say nothing of the many alterations that are inadvertent and do not create human benefits except in the most indirect way: acidification of aquatic systems due to acid precipitation, sedimentation of waters as a byproduct of agriculture or land development, unintentional introduction of invasive exotic species, pollution from city streets, and climate disruption. It is no wonder that, as a group, freshwater aquatic organisms are the single most threatened fauna throughout the world.

The popular notion of how human activities affect rivers focuses on one particular type of impact—pollution, and generally only industrial pollution at that. However, there are many human-induced changes that affect the biological integrity of a river system. The concept of biological integrity of aquatic environments has been defined as "the ability to support and maintain a balanced, integrated, adaptive community of organisms having a species composition, diversity, and functional organization comparable to that of natural habitat of the region" (Angermeier and Karr 1994). Biological integrity of an ecosystem is analogous to the concept of health. For a person to be healthy, certain conditions must be met: all the parts have to be there, working properly. Similarly, for a river to have biological integrity, all the parts—species, physical habitat, flow regime, and water quality—must be in place and working together. If any part is missing or has been modified, biological integrity is lost. Thus, many human impacts on the river biome reduce biological integrity, though they may confer benefits to people.

Stream Restoration

Throughout the industrial countries, there is increasing awareness of the effects of human activity on streams and rivers, beyond a narrow focus on water pollution. This is particularly true in urban areas. A number of national and international organizations focus on stream restoration, and countless local groups with projects focus on their local rivers and stream. The knowledge and technologies for stream restoration have progressed substantially.

Rivers and streams in urban areas seem to stimulate the longing of some people for contact with a more "natural" environment, and the number of urban stream restoration projects continues to grow. Sometimes the effort is to repair the badly eroded channel, plant riparian vegetation, and restore habitat. In other cases, it involves "daylighting" a stream that has been put underground into concrete pipes. Daylighting means exposing the stream again and recreating an engineered channel that approximates a natural channel in appearance and habitat value but still meets stormwater management requirements.

In some cases, the local stream restoration projects—100 yards of stream channel here, 200 yards there—are part of a watershed-wide effort to improve water quality downstream. Just as the water quality at the outlet of a watershed represents an integration of many specific, local conditions throughout the watershed, so too does the improvement of the downstream water quality depend on many local improvements. A case in point in the United States is the Chesapeake Bay Program, a joint program of the U.S. federal government and the state governments of Maryland, Pennsylvania, Virginia to improve the water quality in the Chesapeake Bay. The water-quality problems experienced by the Bay are severe: it is a dying ecosystem, barely hanging on. But they are created by conditions upstream, throughout the watershed. Therefore, the solution lies not in the Bay but in the tens of thousands of miles of tributary streams, agricultural fields, and urban areas in the watershed. The Chesapeake Bay Program provides an umbrella and substantial funding for many stream restoration efforts. Much good work has also been done in the U.S. Pacific Northwest, where stream restoration efforts are focused on improving salmon habitat. Similar large watershed efforts are under way in other parts of the United States, as well as in Europe, Australia, Japan, and elsewhere.

Further Readings

Books

Adler, Robert W. 2007. *Restoring Colorado River Ecosystems: A Troubled Sense of Immensity.* Washington, DC: Island Press.

Campbell, David G. 2005. *A Land of Ghosts: The Braided Lives of People and the Forest in Far Western Amazonia.* Boston: Houghton Mifflin.

Darlington, P. J., Jr. 1982. *Zoogeography.* Malabar, FL: Krieger Publishing.

Leopold, L. 1994. *A View of the River.* Cambridge, MA: Harvard University Press.

Pinkham, R. 2000. *Daylighting: New Life for Buried Streams.* Snowmass, CO: Rocky Mountain Institute. http://www.rmi.org/sitepages/pid172.php.

Postel, Sandra, and Brian Richter. 2003. *Rivers for Life: Managing Water for People and Nature.* Washington, DC: Island Press.

Reisner, Marc. 1986. *Cadillac Desert: The American West and Its Disappearing Water.* New York: Viking Press.

Riley, Ann L. 1998. *Restoring Streams in Cities: A Guide for Planners, Policymakers, and Citizens.* Washington, DC: Island Press.

Internet Sources

Amazonian Fishes and their Habitats. n.d. Based on a multimedia CD collection written by Dr. Peter Henderson, of Pisces Conservation and the University of Oxford, and originally produced by Kathy Jinkings of AquaEco. http://www.amazonian-fish.co.uk.

Center for Watershed Protection. n.d. http://www.cwp.org.

eFish. n.d. "The Virtual Aquarium, Department of Fisheries and Wildlife Sciences, Virginia Tech." http://www.cnr.vt.edu/efish.

Fishbase. n.d. A comprehensive Web-based searchable database of information on fishes. http://www.fishbase.org.

The Nature Conservancy. n.d. "Clinch Valley Program." http://www.nature.org/where wework/northamerica/states/virginia/misc.

Protect Your Waters. n.d. Includes information on the Zebra mussel and other aquatic invaders. http://www.protectyourwaters.net.

Tennessee Wildlife Resources Agency. n.d. "Streams of East Tennessee." http://www .homestead.com/twra4streams/index.html.

Tennessee Wildlife Resources Agency, Region 4 Stream Management. n.d. http://www .homestead.com/twra4streams.

Unionid Gallery, Missouri State University. n.d. http://unionid.missouristate.edu.

U.S. Geological Survey. n.d. "Paddlefish Study Project." http://www.umesc.usgs.gov/ aquatic/fish/paddlefish/main.html.

Wikipedia. n.d. "Amur River," with map and link to World Resources Institute/IUCN land cover map and further information: http://en.wikipedia.org/wiki/Amur.

World Resources Institute/United Nations Environment Programme. n.d. "Watersheds of the World." http://multimedia.wri.org/watersheds_2003/index.html.

World Wildlife Fund. n.d. Varzea forests ecoregion description on World Wildlife's ecoregions. http://www.worldwildlife.org/wildworld/profiles/terrestrial_nt.html.

World Wildlife Fund Conservation Science. n.d. "Freshwater Ecoregions." http://www.worldwildlife.org/science/ecoregions/freshwater.cfm.

Appendix

Selected Plants and Animals of Rivers

Miscellaneous River Biota

Fishes

Upper Klamath sucker	*Chasmistes brevirostris*

Insect Orders

Mayflies	Ephemeroptera
Caddisflies	Trichoptera
Stoneflies	Plecoptera
True flies	Diptera
Beetles	Coleoptera
True bugs	Hemiptera
Alderflies, dobsonflies, fishflies	Megaloptera
Damselflies and dragonflies	Odonata

Insects

Water pennies	Family Psephenidae
Hydropsychid caddisflies	Order Trichoptera; Family Hydropsychidae

The Amazon River

Plants (Main Channel)

Water paspalum	*Paspalum repens*
Aleman grass	*Echinochloa polystachya*
Eared watermoss	*Salvinia auriculata*

Plants (Channel Bars and Low-Lying River Shores)

Water paspalum	*Paspalum repens*
Aleman grass	*Echinochloa polystachya*
Mexican crowngrass	*Paspalum fasciculatum*

Plants (Low-Lying Flats, Floodplain Lake Beds)

West Indian marsh grass	*Hymenachne amplexicaulis*
Burrhead sedge	*Scirpus cubensis*

Plants (Varzea)

Lythraceae	*Adenaria floribunda*
Iporuru	*Alchornea castanaefolia*
Willow	*Salix martiana*
Early successional tree (no common name)	*Annona hypoglauca*
Spiny palm	*Astrocaryum jauari*
Early successional tree (no common name)	*Cecropia latiloba*

Invertebrates

Amazon river prawn	*Macrobrachium amazonicum*
Apple snail	*Pomacea lineata*
Mayfly	*Asthenopus curtus*

Birds

Snail-kite	*Rostrhamus sociabilis*

Fishes

Black acara	*Cichlasoma bimaculatum*
Tambaqui	*Colossoma macropomum*
Piraiba	*Brachyplatystoma filmentosum*
Candiru	*Vandellia cirrhosa*
Electric eel	*Electrophorus electricus*
Oscar fish	*Astronotus ocellatus*
Aripaima	*Arapaima gigas*

Other vertebrates

Black caiman	*Melanosuchus niger*
Amazon river dolphin	*Inia geoffrensis*
Gray dolphin	*Sotalia fluviatilis fluviatilis*

(*Continued*)

Amazon manatee	*Trichechus inunguis*
Giant otter	*Pteronura brasiliensis*
Amazonian river otter	*Lutra enudris*
Giant Amazonian river turtle	*Podocnemis expansa*
Matamata	*Chelus fimbriatus*
Anaconda	*Eunectes murinu*

The Amur River

Plants

Diatom	*Aulacosira islandica*

Invertebrates

Mussels	*Middendorffinaia mongolica, Dahurinaia dahurica*

Fishes

Amur grayling	*Thymallus arcticus*
Lenok	*Brachymystax savinovi*
Mongolian taimen	*Hucho taimen*
Amur sturgeon	*Acipenser schrenckii*

Fishes (Floodplain Lakes)

Carp	*Cyprinus carpio haematopterus*
Amur pike	*Esox reichertii*
Amur catfish	*Parasilurus asotus*
Black Amur bream	*Megalobrama terminalis*
Chinese perch	*Siniperca chuatsi*
Snakehead	*Channa argus warpachowskii*

Fishes (Migratory)

Pacific salmon	*Oncorynchus gorbuscha, O. tshawytscha*
Lamprey	*Lampetra japonica*
European smelt	*Osmerus eperlanus*

Rivers of Southern Appalachia: The Upper Tennessee

Plants (Riparian)

Water willow	*Justicia americana*
Red maple	*Acer rubrum*
Buttonbush	*Cephalanthus*

Cottonwood	*Populus deltoids*
Black gum	*Nyssa sylvatica*
American sycamore	*Platanus occidentalis*
Black willow	*Salix nigra*

Invertebrates

Mussels	See Table 2.2.
Rusty crayfish	*Orconectes rusticus*
Obey crayfish	*Cambarus obeyensis*

Fishes

Tennessee shiner	*Notropis leuciodus*
Mirror shiner	*Notropis spectrunculus*
Flat bullhead	*Ameiurus platycephalus*
River chub	*Nocomis micropogon*
Fantail darter	*Etheostoma flabellare*
Swannanoa darter	*Etheostoma swannanoa*
Tangerine darter	*Percina aurantiaca*
Striped shiner	*Luxilus chrysocephalus*
Telescope shiner	*Notropis telescopus*
Bluntnose minnow	*Pimephales notatus*
Banded sculpin	*Cottus carolinae*
Golden redhorse	*Moxostoma erythrurum*
Greenside darter	*Etheostoma blennioides*
Redline darter	*Etheostoma rufilineatum*
Stoneroller	*Campostoma anomalum*
Whitetail shiner	*Notropis galacturus*
Striped shiner	*Luxilus chrysocephalus*
Snubnose darter	*Etheostoma simoterum*
Paddlefish	*Polyodon spathula*
Lake sturgeon	*Acipenser fulvescens*
Snail darter	*Percina tanasi*

Other Vertebrates

Shovelnose salamander	*Leurognathus marmoratus*
Hellbender	*Cryptobranchus alleganiensis*

Rivers of Southern Appalachia: The New

Fishes

Green sunfish	*Lepomis cyanellus*
American eel	*Anguilla rostrata*

(Continued)

Bigmouth chub (endemic)	*Nocomis platyrhunchus*
Kanawha minnow (endemic)	*Phenacobius teretulus*
New River shiner (endemic)	*Notropis scabriceps*
Bluestone sculpin (endemic)	*Cottus* spp.
Cave sculpin (endemic)	*Cottus* spp.
Appalachia darter (endemic)	*Percina gymnocephala*
Kanawha darter (endemic)	*Etheostoma kanawhae*
Candy darter (endemic)	*Etheostoma osburni*

Fishes (Introduced)

Largemouth bass	*Micropterus salmoides*
Smallmouth bass	*Micropterus dolomieu*
Bowfin	*Amia calvia*
Alewife	*Alosa pseudoharengus*
Golden redhorse	*Moxostoma erythrurum*

3

Wetlands

Introduction to the Wetland Biome

In 1971, at a meeting in Ramsar, Iran, most of the world's industrial countries and a large number of developing countries agreed that wetlands were valuable environmental assets that should be protected. By 2006, 153 countries had agreed to the principles of the Ramsar Convention, as it is now known; and many countries, including the United States, now have national laws protecting wetlands as well. This international consensus that wetlands are worthy of protection represents a complete turnaround from the attitude toward wetlands most people held for hundreds of years: that wetlands were, at best, wild places, uninhabited obstacles to human progress; at worst, they were waste places harboring disease. Before it was discovered that microorganisms caused infectious diseases, many scientists believed that "miasmas"—poisonous airs arising in swamps—caused the terrible diseases (typhoid and cholera among others) that killed millions of people every year.

It must be acknowledged that the change is incomplete. While governments may be well-intentioned toward wetlands, many in society cling to the old view of wetlands. More important, those who own wetlands may acknowledge that wetlands are beneficial for society as a whole, but the ways in which wetlands are valuable for society don't pay the rent. Therefore, in the absence of government regulation, many property owners choose to destroy wetlands and use the land for agriculture or urban development. As a result, the loss of wetlands continues, perhaps at a slower pace than before the 1970s, but it continues.

Incomplete though it may be, the turnaround regarding wetlands is primarily due to scientific recognition of the socially valued functions of wetlands (see Table 3.1). Wetlands regulate and prevent downstream flooding; they remove nutrients and sediment—potential pollutants—from waters; they provide habitat for a large number of plants and animals; they provide recreational opportunities; and they give pleasure to those who have learned to appreciate their beauty.

The U.S. Fish and Wildlife Service (USFWS), charged with major responsibilities regarding wetlands protection in the United States, defines wetlands as "lands transitional between terrestrial and aquatic systems where the water table is usually at or near the surface or the land is covered by shallow water" (Cowardin, Golet, and LaRoe 1979). Three characteristic elements identify wetlands: the presence of water-loving (hydrophilic) *plants*; *soils* that show the telltale signs of being saturated with water all or much of the time (hydric soils); and *water*, more specifically, water level in relation to the land surface. These three wetland elements are used by U.S. government agencies for wetland identification in the field. Wetland identification is important from a practical point of view, because if an area is identified as a wetland, then certain restrictions on the use of that land may apply. These restrictions may have significant implications for the value of the land.

Wetlands classification systems have been developed and are used to facilitate the work of wetlands scientists, biologists and ecologists, environmental planners, and land managers. Wetlands are classified according to one or more of the following factors: hydrology, including hydrodynamics and hydroperiod (see below), as well as water source (which determines water quality and trophic status to a large extent); saltwater versus freshwater; position in the landscape (geomorphology); and vegetation, which integrates the other factors but also includes the effects of climate and biogeography.

Wetland Characteristics

Water

Wetlands at the most basic level are just what the word implies: wet lands. How wet and for how long determine which plants and animals will be found in a wetland. The pattern of rises and falls in water level over time in a particular wetland is called its hydroperiod, the "signature" of water on that particular land area. The graphic representation of the hydroperiod is similar to the hydrograph of a river, which relates river discharge (and thus water level) to time, but typically the hydroperiod's timescale is longer than those used for river hydrographs (see Figure 3.1).

The water level's ups and downs could be caused by surface and near-surface runoff of precipitation (in basin-shaped wetlands), by surface or subsurface lateral flows caused by rising and falling water levels of nearby lakes or rivers (in fringe, riverine, and tidal wetlands), or by fluctuations in groundwater levels.

Hydroperiods are sometimes categorized according to the length of surface inundation. In nontidal wetlands (wetlands not influenced by ocean tides),

Table 3.1 Wetland Functions and Values

FUNCTION	EFFECT	SOCIETAL VALUE
Hydrologic		
Short-term surface water storage	Reduced downstream flood peaks	Reduced property and crop damage from floodwaters
Long-term surface water storage	Maintenance of stream flows, seasonal stream flow moderation	Maintenance of fish habitat during dry periods
Maintenance of high water tables	Maintenance of hydrophytic plants, groundwater for tree and crop growth	Maintenance of biodiversity, increased timber and crop production
Biogeochemical		
Transformation and cycling of elements	Maintenance of nutrient stocks within wetland, production of dissolved and partially decayed organic matter	Timber production, food for fish and shellfish downstream, support of recreational and commercial fishing
Retention, removal of dissolved substances	Reduced transport of nutrients and pesticides downstream	Maintenance of water quality; safer drinking water
Accumulation of peat	Retention of nutrients, carbon, metals, other substances	Maintenance of water quality, reduction of global warming
Accumulation/retention of inorganic sediment	Retention of sediment and attached pesticides, phosphate and other nutrients	Maintenance of water quality, clear water, high-quality fish populations in streams
Habitat and Food Web Support		
Maintenance of characteristic plant communities	Food, refuge, and nesting cover for wildlife; spawning, refuge and nursery habitat for fish and shellfish; food for humans	Support for waterfowl and other wild game, furbearers, uncommon and rare and endangered species, fish, and shellfish; recreational and commercial hunting, fishing and bird watching
Maintenance of characteristic energy flow	Support for populations of vertebrates and invertebrates	Maintenance of biodiversity, bird watching, aesthetics

Sources: After U.S. Army Corps of Engineers, http://www.usace.army.mil/cw/cecwo/reg/wet-f-v.htm; and NRC 1995.

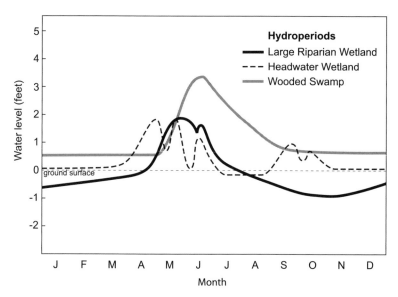

Figure 3.1 Hydroperiods of three types of North American wetlands. *(Illustration by Jeff Dixon. Adapted from Welsch et al. 1995.)*

hydroperiods may be *permanent* (flooded all year every year); *intermittent* (flooded throughout the year except in very dry years); *seasonal* (flooded in the growing season every year); *saturated* (standing water is rarely present but the soil is saturated every year at least during the growing season); and *temporary* (inundated for a relatively brief period during the growing season, after which soil is not saturated). In tidal wetlands, including freshwater tidal wetlands, hydroperiods may be categorized as subtidal (flooded with tidal water even at low tide); irregularly exposed (the substrate is exposed by the tides less frequently than daily, for example only during unusually low tides); regularly flooded (flooded then exposed at least once daily); and irregularly flooded (flooded less often than daily).

In nontidal wetlands the hydroperiod depends primarily on precipitation and evapotranspiration. Vernal pool wetlands in eastern North America have water above the ground during the winter and spring, when evapotranspiration rates are low. During the warmer months, evapotranspiration rates increase and the water table retreats below the ground. The hydroperiods of riparian wetlands (wetlands associated with rivers) are connected with the ups and downs of their rivers. The flooded forests of the Amazon River in Brazil (see Chapter 2) flood regularly once a year for several months. The water level rises dramatically from below the ground surface to many feet above. A period of inundation of at least a week during the growing season is about the minimum for the formation of a wetland's characteristic soils.

The changes in water level can be analyzed in terms of inputs and outputs of water to a wetland—in other words, the water budget. Possible inputs or sources of water—which may be different for different types of wetlands—include ground-water discharge into the wetland, direct precipitation onto the wetland, and surface flow into the wetland. Possible outputs of water include percolation back into groundwater (groundwater recharge), evapotranspiration, and surface flow out of the wetland. With water, nutrients, sediment, organic material, dissolved solids, and organisms may come (or go). For example, floodplain wetlands regularly receive large inputs of sediment, nutrients, and organic materials from flooding rivers. But if a wetland's water leaves solely via evapotranspiration, nutrients, sediment, organic material, and dissolved solids will be left behind. This may result in a buildup of salts that eventually will turn a freshwater into a saline wetland. The buildup of organic material and sediment can change a wetland from one type into another over time, perhaps lengthening the period of saturation as soil pore spaces are clogged with fine sediments and organic material.

The water level that determines a wetland's degree of wetness is called the water table. The water table is the top of the subsurface zone of saturated soil, the zone in which pore spaces between soil particles are filled with water instead of air. Nonwetland areas might have a water table anywhere from a few yards to hundreds of feet below the ground surface. In most wetland areas, the water table is close to the ground surface or even above it, resulting in ponding of water on the surface. Ombrotrophic wetlands—wetlands that are fed only by local precipitation (rain and snow)—are the exception. Such wetlands, to exist at all, must be on soils that are relatively impermeable; that is, water cannot pass through them easily (instead of infiltrating, it collects in surface depressions and over time a wetland forms).

A wetland's trophic status is closely related to its source of water. Wetlands associated with river systems and wetlands fed by groundwater tend to have higher levels of key plant nutrients than those fed only by rainfall, although there are exceptions—for example, the flooded forests of the nutrient-poor black-water and clear-water tributaries of the Amazon.

Soils

Wetlands are characterized by soils known as hydric soils. The U.S. Department of Agriculture, Natural Resources Conservation Service, defines hydric soils as soils "that formed under conditions of saturation, flooding or ponding long enough during the growing season to develop anaerobic conditions in the upper part." These "anaerobic conditions" refer to a lack of oxygen caused by saturation with water. Most nonhydric soils are only briefly and occasionally saturated and therefore are aerobic. Plant roots in such soils use the readily available oxygen to support their growth. Organic material (dead plant roots, leaves, stems, and soil organisms) in aerobic soils decomposes readily, as oxygen is available to support the activities of decomposer organisms in the soil.

In anaerobic soils, decomposition of organic material occurs slowly; under some conditions (low temperatures, for example), decomposition may be so slow that the annual addition of organic material is greater than the annual removal of organic material by decomposition. The upper soils in peat wetlands are made up of almost 100 percent organic material, and individual plant parts can easily be identified even after centuries.

Wetland soils are classified as either *organic* or *mineral*. Organic soils have a high proportion (greater than a third) of organic material. Compared with mineral soils, their ability to hold water, like a sponge, is high. They are also quite permeable, meaning that water passes through them readily. Therefore, wetlands with organic soils can form only when there is a layer of relatively impermeable material beneath the hydric soils to keep water from percolating down into the groundwater system.

Organic soils tend to be acidic and nutrient poor. They form under anaerobic conditions resulting from frequent or continuous saturation in which the rate of decomposition is low, often because of low temperatures and short growing seasons. Organic wetland soils are divided by soil scientists into three groups, the fibrists (peat), the hemists (peaty muck), and the saprists (muck) (see Table 3.2).

Mineral soils have less than one-third organic content. They hold less water than organic soils and also have relatively low permeability. Because water cannot move through them readily, anoxic conditions develop fairly rapidly upon saturation and persist. Mineral soils tend to be neutral rather than acidic. Nutrient availability is relatively high. Mineral wetland soils have identifiable characteristics,

Table 3.2 Hydric Organic Soil Characteristics

Organic Soil Type	Color	Characteristics
Fibrists (peat soils); Soil Order Histosol	Brown to black	Wet histosols in which organic materials are only slightly decomposed. Plant material recognizable; low bulk density; greatest ability to hold water like a sponge (porosity)
Saprists (muck soils); Soil Order Histosol	Black	Wet histosols in which organic materials are well decomposed. Will stain fingers when moist; runny when wet; may have "rotten" odor; few if any plant fibers recognizable; higher bulk density; lower porosity
Hemists (mucky peats); Soil Order Histosol	Brown to black	Wet histosols in which organic materials are moderately decomposed; all characteristics are intermediate between those of the Fibrists and Saprists, the Hemists representing a state of decomposition greater than Fibrists but less complete than Saprists

including a dull gray background coloration typically mottled with reddish areas. Such characteristics are created by biochemical transformations resulting from the low-oxygen conditions associated with saturation. Iron, manganese, sulfur, and carbon are involved in these transformations, which are collectively referred to as oxidation and reduction or redox reactions. Under low-oxygen conditions, soil microbes "reduce" iron oxides, making the iron easily dissolved so that it moves with the water. It is typically leached from (washed out of) some soil areas but may become oxidized again in areas with higher oxygen levels, or when the soil is exposed to air. Higher oxygen levels under saturated conditions occur around the roots of plants specially adapted to diffuse oxygen to their roots. An orange- or rust-colored accumulation of iron oxide in such areas, together with a gray appearance in the soil areas from which the iron was leached, lends the soil its mottled appearance.

Vegetation

Wetlands form in virtually any geographic setting whose climate allows plants to grow. Given a location with the appropriate conditions of soils and hydrology, a wetland ecosystem will develop. Locations in which palustrine (inland, freshwater) wetlands are found include the floodplains of rivers, particular in a river's lower reaches, and poorly drained basins or depressions. Many large wetland complexes—such as the Everglades in Florida; the Mackenzie River Basin in Canada; and the Pantanal in Brazil, Bolivia, and Paraguay, to name just a few—include several types of wetlands.

Plants that are found in wetlands are either obligate (plants that only grow in a wetland environment) or facultative (plants that can grow in wetlands but can also thrive in other environments). Obligate wetland plants occur in wetlands 99 percent of the time and only 1 percent are found outside wetlands, whereas facultative wetland plants are found in wetlands 67–99 percent of the time. Obligate wetland plants are used as wetlands indicators—plants whose presence indicates the existence of wetland conditions with a high degree of confidence. Wetland plants have a number of adaptations that enable them to live in wetlands; these adaptations are discussed in the section "Life in a Wetland," below. It is not possible to present a general list of "typical" wetland plants, as the plant communities differ considerably depending on the type of wetland and other factors. For example, marshes are dominated by herbaceous emergent vegetation adapted to conditions of frequent or continuous inundation. They differ from swamps, which are dominated by woody shrubs and trees, and from bogs, which are often dominated by mosses. Typical plants are presented in the descriptions of the different wetland types later in this chapter.

Extent and Geographic Distribution of Wetlands

Wetlands are found in every geographic setting, on every continent except Antarctica. There are wetlands in the desert, wetlands in the Arctic, wetlands in the rainforest, and wetlands in the most densely populated regions. Estimates of the global

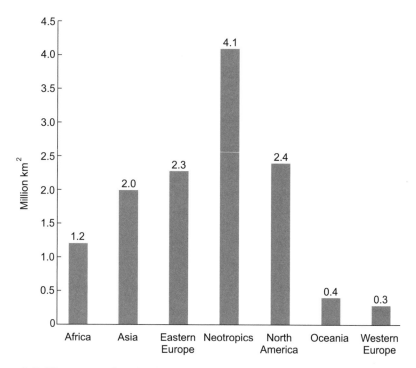

Figure 3.2 The extent of wetlands around the world. *(Illustration by Jeff Dixon. Adapted from Finlayson et al. 1999.)*

extent of wetlands vary greatly, from 3.3 to about 8 million mi^2 (5.3 to 12.8 million km^2), which equates to between 3.5 and 8.5 percent of a total land surface area of about 93 million mi^2 (150 million km^2). At first glance, one might think that the global extent of wetlands was a well-known fact, but this is not the case. While it is true that the entire surface of the Earth can be and is viewed, photographed, and analyzed via Earth-orbiting satellites, it is not easy to distinguish precisely wetlands from other lands based on remotely sensed data. National and regional geospatial data sets and wetlands inventories are of inconsistent quality and definitions vary. Furthermore, wetlands are a moving target. Seasonal wetlands dry up; El Niño and other periodic phenomena cause wetlands to grow and shrink over time. People alter wetlands everywhere, filling them in, draining them, paving over them, restoring them. Figure 3.2 shows the area of wetlands in different world regions.

Wetlands Trends

Given the uncertainty regarding the extent of the world's wetlands, it is not surprising that there are no precise figures on the loss of wetlands worldwide either. The United States has relatively well-documented figures on wetlands and wetlands loss. As of 1997, the USFWS estimated that there were 106 million acres of wetlands in the lower 48 states, including about 101 million acres (95 percent) of freshwater wetlands and 5 million acres (5 percent) of saltwater wetlands. This total represents less than 50 percent of the estimated 221 million acres of wetlands in the lower 48 states in the 1600s. Recent studies by USFWS have shown a leveling off of freshwater wetland losses in the United States, but this optimistic overall trend masks continued losses of marshes and swamps, offset in the statistics by increases in manmade farm ponds.

While precise data are lacking, it is reasonable to suppose that human conversion of wetlands to agricultural and urban land uses has reduced the world's wetlands area by 50 percent or more over several centuries, with the highest percentages lost in developed regions. The ecological functions and values of wetlands, however, can be degraded without wholesale conversion. Pollution, over-harvesting of plants or animals, and the introduction of exotic invasive species can degrade wetland ecosystems even though they remain wetlands. National-level area statistics of wetlands do not incorporate such qualitative considerations.

Life in a Wetland

Wetlands as living environments present many challenges to organisms, necessitating various adaptations. The main challenges are conditions of low or no oxygen (anaerobic conditions); fluctuating water levels so that, in the extreme, plants, animals, and microorganisms must be able to survive extended inundation and extended periods of dry conditions; the presence of phytotoxic concentrations of substances resulting from biochemical transformations caused by anaerobic conditions; and the presence of salt in marine or estuarine wetlands.

Microorganisms

The adaptations of bacteria and protists to freshwater wetland conditions are complex and various. They primarily involve adaptations of cellular biochemistry that allow the organisms to respire and engage in cellular metabolism without using oxygen. Anaerobic bacteria have developed mechanisms at the subcellular level that allow them to deal with the toxic end products—such as lactic acid—of anaerobic metabolism, either by detoxifying these substances or expelling them. They are able to use reduced organic compounds in wetlands soils as a source of energy and also can use inorganic soil elements in place of oxygen as electron receptors. Facultative anaerobic bacteria are able to switch from using oxygen as an electron receptor to using these other elements. Obligate anaerobic bacteria use sulfate and are responsible for the production of hydrogen sulfide, the gas whose odor is sometimes associated with wetlands (the "rotten egg" smell).

Wetland Plants

The presence of plants specifically adapted to living in wetlands is one of the characteristics by which scientists identify wetlands. At the cellular level, physiological and metabolic changes are similar to those seen in one-celled organisms. But plants are also able to grow special structures and change their morphology (growthform) in ways that allow them to survive in low-oxygen environments. For emergent wetland plants—plants that are rooted in saturated substrates but have most of their growth above water—the problem is confined to their roots, which must have oxygen to function. If the roots do not function, they cannot pull water or nutrients up into the leaves, and the plants die.

The development of aerenchyma, spongy or cork-like tissue that consists of relatively large intercellular voids or spaces, allows oxygenated air to diffuse to the roots from above-water parts of the plant. Some plants, such as some alders (for example, the speckled alder and European alder), develop aerenchyma only in response to anaerobic conditions; if they are growing in an upland environment, they will not do so. Often special structures (for example, pneumatophores or "air roots") contain the aerenchyma.

The transfer of oxygen to the roots of wetland plants results in excess oxygen being given off by the roots and creating an aerobic soil microenvironment around the roots. This well-oxygenated film permits the development of mycorrhizae, symbiotic associations between plant root hairs and certain fungi, that allow for more effective root functioning on the part of the plant and are an adaptation allowing plants to live in wetlands environments. The oxygenated zone is responsible for the reddish-colored deposits of iron oxide that characterize hydric soils and allows for the oxidization of sulfides and reduced metal ions that renders them nontoxic to plants.

Low-oxygen conditions can stimulate the production of the pneumatophores or air roots with aerenchyma, as mentioned above. The "knees" of the bald cypress (see Plate V) are examples of such morphological adaptations. Elongated roots called "prop roots" also may have pneumatophores on them; these are commonly seen on wetland plants such as the red mangrove.

In some common emergent, floating-leaved, and woody wetland plants, air is actually forced down to the roots under pressure. Plants for which this has been demonstrated include the water lily (family Nymphaeaceae), the lotus (genus *Nelumbo*), the common reed, the southern cattail, and the common or European alder. Such pressurization is induced by temperature and humidity differentials, and the amount of pressure the plants can generate is apparently related to the depth of their roots. Inundation and resulting low-oxygen conditions stimulate some plants to elongate their stems quickly to keep their leaves above water. Such an adaptation is frequent among plants inhabiting wetlands with a long, slow, predictable or seasonal period of inundation, such as those of the Amazon floodplain forests. Rice and bald cypress are among the plants in which rapid stem elongation has been observed.

Wetland plants have developed what might be termed behavioral adaptations. In wetlands with long, predictable hydroperiods, plants often time their seed production to coincide with the receding water or, in some cases, to coincide with rising water. Some have seeds that float and resist waterlogging. Others have seeds that can survive long periods under water and germinate during the odd drought year. Yet others produce seeds that can germinate under water, or seeds that germinate while still attached to the tree so they do not have to drop into the water until the floods recede. In the Amazonian *varzea* (see Chapter 2), some trees deal with the prolonged periods of inundation by dropping their leaves and going dormant.

Sphagnum, a genus of mosses containing about 150 species, and the dominant among peat mosses, take a different tack from most wetland plants in adapting to conditions of waterlogging. The plants maintain a waterlogged condition as their preferred state; that is, they hold a great deal of water in their tissues. Sphagnum also has the unusual ability to acidify its surroundings, which gives it a competitive advantage as most other plants cannot survive in acidified waters. The level of acidity maintained by sphagnum reduces bacterial activity, slowing decomposition and leading to the accumulation of peat in peat bogs.

In general, plant adaptations are optimized for a specific set of circumstances, for example, length of flood period, depth of water, or water chemistry conditions. This fact results in often-observed vegetation zones arranged along gradients of elevation, salinity, water depth, or nutrient supply.

Wetland Animals

The enormous variety—geomorphic, hydrologic, ecoregional—of habitats included in the term "wetlands" precludes the easy listing of wetland species. Floodplain or riverine and fringe wetlands share many animals with the larger water bodies with which they merge periodically. In the Amazon, fish species that during low water live either in the river channel or in floodplain lakes fan out into the flooded forests during the lengthy periods of inundation. Floodplain wetlands are used by riverine fishes and other riverine species (for example, otters) during the flood period; the ecological significance of this fact is highlighted in the Flood-Pulse Concept (FPC) (see Chapter 2). Wetlands not connected to rivers and lakes by surface waters often lack fish, but may nonetheless share more mobile and cosmopolitan (widely distributed) reptiles, amphibians, and invertebrates (particularly the larval stage of flying insects) with those water bodies. The sections below introduce selected examples of common wetlands animals.

Invertebrates. In wetlands, insects are common and include many of the taxa that live in other freshwater environments (see Chapter 1). Their aquatic larvae are important links in detrital food chains in marshes and swamps, feeding fish, amphibians, and birds. Mosquitoes (members of order Diptera) and dragonflies and damselflies (order Odonata) are well-known wetland insects. Species of both taxa have adapted to all kinds of wetlands and are widely distributed.

Vertebrates. Amphibians are, perhaps more than any other group of animals, associated in the popular mind with wetlands. Wetland examples include frogs, toads, salamanders, and newts. The American bullfrog adds a familiar voice to the night-time choir in North American wetlands.

Reptiles are rightly associated with wetlands in the popular imagination. While relatively few reptiles are obliged to live in water, many spend much or most of their lives in freshwater environments. Members of the order Crocodilia inhabit wetlands, lakes, and rivers in Africa, the Americas, southern and east Asia, and Australasia. Today's crocodilians—crocodiles, alligators, and caimans—are not much changed since their ancestors in the Cretaceous period (about 84 million years ago). They are omnivorous and some species may grow quite large. The cai-man known as the jacare is one of the most highly visible and ubiquitous of the Pantanal's animals. The American alligator, another member of the family Alliga-toridae, occurs throughout the southeastern United States. Turtles and snakes are often found in wetlands, particularly in temperate and tropical regions.

Fishes are present in most freshwater wetlands. Fringe and riverine wetlands share many species with the lakes and rivers to which they are connected, and often serve as "nursery habitat" for fish species that live in open water as adults. Isolated wetlands, such as prairie potholes, may not have any fish or only introduced fish. Bogs, particularly those with low pH, tend to be challenging environments for fish. Yet even here there are fish. The world's smallest fish, a member of the carp family, was discovered in a peat bog on the island of Sumatra. Only 0.31 in (7.9 mm) long, it manages to live in water with a pH of about 3, so acidic that most fish could not survive in it, much less reproduce.

Wetlands are home to many birds; their invertebrate populations provide a food source of great importance even to terrestrial birds. The association of water-fowl with wetlands is well founded; ducks, geese, and coots all prefer to inhabit wetlands. Certain birds are emblematic of particular wetlands. For example, the Jabiru Stork has come to symbolize the Pantanal, while the Caribbean Flamingo is associated with the Florida Everglades.

The abundance of food in wetlands, ranging from grasses to insects, amphib-ians, and nesting birds, attracts many terrestrial mammals. Most mammals are fre-quent visitors to wetlands, but relatively few are lifelong inhabitants. Typical residents include otters, muskrat, nutria, beaver, mink, raccoon, swamp rabbit, marsh rice rat, hippopotamus, and water buffalo. Many wetland mammals are her-bivores or omnivores. However, certain carnivores lead semiaquatic lives in and around wetlands, including the South American jaguar and the Asian fishing cat.

Ecological Processes in Wetlands

For decades it was thought that wetlands were a transitional ecological commu-nity, part of a successional process but not the end point (climax) of that process. Marshes were seen as an early sere in a process that started with open water and inexorably led to a forest community. The buildup of sediment and organic mate-rial, it was supposed, ultimately would fill in a wetland. Vegetation zones, obvious in many marshes, were seen as evidence of succession. Some marshes do fit that description, but others have persisted with similar vegetation for hundreds of years—in other words, the marsh is the "climax" or perhaps "subclimax" community.

It appears that a wetland's type and its biotic community are the result of a number of factors generally unique to a particular site or region. These factors include the wetland's soils, hydroperiod, water source, seed bank and available plant and animal propagules, climate, and (not least) its unique history. Of particular importance in a wetland's history is its record of disturbance—disruption by fire, flood, drought, hurricane, pollution, logging, flooding with saltwater, or agricultural use. The vegetation zones seem to be related to environmental gradients of (primarily) inundation and saturation, not succession. Thus, there is no reason to suppose that a freshwater marsh in a forested region is necessarily on its way to becoming forest, whether wetland or upland. Indeed, the interactions of some wetland environments with their plants and microorganisms produce conditions that preclude the establishment of other types of plants. For example, as sphagnum moss becomes established, it acidifies its environment, making it inhospitable to most other plants.

Most wetlands are relatively stable features of the landscape, given a stable climate regime. But "relatively" here refers to human timescales; on a timescale of tens of thousands of years, all bets are off as climate itself varies considerably over tens of millennia.

Types of Freshwater Wetlands

Tidal Freshwater Marshes

A marsh is a type of wetland dominated by herbaceous emergent vegetation adapted to the prevailing conditions of frequent or continuous inundation. Tidal freshwater marshes are characterized by hydroperiods driven by tidal influences. They occupy a distinct place in the landscape, generally along the margins of rivers or in river deltas that are far enough from oceans that they are freshwater systems but are still affected by tides. They are usually inland from saltwater marshes, high up in estuaries. On the U.S. Atlantic coast, many are near cities, since many of the major cities were built just below the fall line of major rivers, the point at which rivers go from tidal to nontidal. In urban areas such as Philadelphia, Richmond, and Washington, D.C., human impacts on these marshes are longstanding and in some cases severe. Because of their generally flat topography and rich soils, many tidal marshes in both North America and Europe have long since been diked, drained, and turned into agricultural land.

Tidal marshes may be either saltwater, freshwater, or somewhere in between along a gradient of salinity (see Figure 3.3). Salinity is generally measured as parts per thousand (ppt), and ocean water has a salinity of 35 ppt (plus or minus about 2 ppt due to natural variations), or about 3.5 percent. A unit now used to measure salinity is the practical salinity unit (PSU), according to which ocean water is

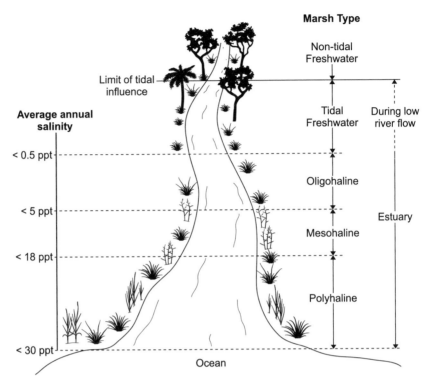

Figure 3.3 Location of tidal freshwater marshes along a gradient of declining salinity in the lower reaches of a river. *(Illustration by Jeff Dixon. Adapted from Odum et al. 1984.)*

35 (no units are stated). Freshwater's salinity is less than 0.5. Water above 0.5 but less than 5 is considered oligohaline, 5–18 is mesohaline, 18–30 is polyhaline, and greater than 30 is euhaline.

Along a typical river-estuary complex, the different salinity zones shift upstream or downstream seasonally or from year to year. In a drought year, when river flows are low, the more saline waters move up the river; in a wet year, or wet season, fresher water pushes downstream. Too salty, and many obligate freshwater marsh plants will die, to be replaced by more salt-tolerant species. Too little salt, and saltwater-tolerant species will be outcompeted by obligate freshwater species. There is little overlap in terms of plant species composition between salt- and freshwater tidal marshes. Figure 3.3 shows the salinity zones in a typical river along which saltwater and freshwater tidal marshes might be found.

The plant communities that compose tidal marshes may shift either toward or away from more salt-tolerant plant species on timescales related to changes in the salinity zones. Rising sea levels, both observed and predicted as climate change intensifies, will push salinity higher up into estuaries, and doubtless change some freshwater tidal marshes to saltwater communities.

Three conditions characterize freshwater tidal wetlands: (1) a source of tidal action; (2) flat topography or very low gradient from the freshwater source; and (3) a supply of fresh water sufficient to exclude salt water all of the time. Tidal freshwater marshes occur on all continents, generally along coastlines in the mid- to high-latitudes. They occur infrequently in the Tropics, where tidal wetland systems tend to be forested. In the United States, they occur along the Atlantic coast (405,000 acres, or 164,000 ha) and the Gulf Coast (946,000 acres, or 383,000 ha), mostly in Louisiana.

The hydrodynamics of tidal freshwater marshes is plain: water flows in and out of these wetlands from a tidally connected surface water body. It is typical for tidal freshwater marshes to have a network of streams that distribute water through the marsh as it rises and drain it as it recedes. The flow of water in and out of the marsh, and the corresponding rise and fall of water levels, is controlled by the tides. Regular ocean tides, caused by the gravitational pull of the moon and, to a lesser extent, of the sun, usually occur twice daily—that is, two high tides and two low tides, every 24 hours and 48 minutes (a period that is determined by both the lunar orbit around the Earth and the Earth's rotation). The amplitude of the tide—how much the water goes up and down—is determined by a number of factors, including the positions of the moon and sun and the topography and bathymetry of the coastline. Amplitude is often increased by the funnel shape of river channels where they enter the sea. As the tide moves upstream, the constriction of the narrowing river may exaggerate the water elevation. Well upstream of the coast, the ups and downs of water that tidal wetlands experience may be dramatic.

The other type of tide, wind tides, may be a phenomenon not only of coastal rivers and estuaries but lakes as well. A wind tide results when a strong, steady wind exerts force on a body of water, essentially pushing it up against the shore or, alternatively, pushing it away from the shore. Unlike the regularity of lunar tides, the timing, duration, and amplitude of inundation or exposure to the air caused by wind tides are usually unpredictable.

As the tide rises, water inundates more and more of the wetland area. The lowest area, known as the low marsh, is inundated to a greater depth and (more important) for a longer time than the high marsh, which is farther upslope. These two zones typically have different dominant vegetation, but as is usually the case in nature, there is less a hard dividing line than a gradual transition between them. Zonation of vegetation is not as pronounced as in salt marshes but, in general, plant species diversity is higher.

In North American tidal freshwater marshes, the low marsh is characterized by low, broadleafed perennial plants, relatively low production, and a relatively high proportion of plant biomass in roots and rhizomes. The high marsh, in contrast, is dominated by perennial grasses, sedges and similar tall plants, with relatively high production and generally greater plant species diversity.

A study of tidal freshwater marshes along small New Jersey rivers and creeks draining into the Delaware estuary revealed low marsh areas characterized by spatterdock, pickerelweed, and arrow arum, sometimes interspersed with wild rice. The low marsh areas were typically exposed only during low tide. High marsh areas were characterized by calamus, halberdleaf tearthumb, jewelweed, relatively pure stands of wild rice, cattail, and various grasses. A mid-marsh zone was distinguished by the presence of marsh smartweed and tidalmarsh amaranth. In this study, a significant increase in low marsh areas at the expense of mid and high marsh areas was observed over two decades, perhaps because of the rising sea level.

A quite different type of tidal freshwater marsh featuring floating mats of vegetation is found along the northern coast of the Gulf of Mexico in the southern United States. Floating mat marshes are not uncommon around the world, including the *varzea* floodplains of the Amazon and its tributaries (see Chapter 2), but most are not tidal. In Louisiana, on the Mississippi River delta, floating marshes are dominated by maidencane, often in association with Olney threesquare, common cattail, and giant bulrush. Such marshes are thick floating carpets with a diversity of plants, including vines and ferns among the dominant plants; this carpet rises and falls with the tides.

The Gulf Coast is geomorphically active, with new deltas forming as sediments are deposited from the Mississippi watershed. On these new deltas, freshwater tidal marshes also occur, dominated by willows on the higher ground and by sedges, arrowheads, cattails, and other annuals and perennials in the lower elevations. Animal communities of freshwater tidal marshes show a relatively low diversity of invertebrates, other than insects, compared with saltwater tidal marshes that are dominated by marine and estuarine organisms—clams, crabs, shrimp, and the like. Food chains are predominately detritus-based, with small organisms such as nematodes and macroinvertebrates like insect larvae, freshwater snails, oligochaete worms, freshwater shrimp, and amphipods playing a central role in processing detritus and making its food energy available to higher trophic levels.

Freshwater tidal marshes may have a relatively high diversity of reptiles and amphibians, birds, and fur-bearing mammals compared with saltwater marshes nearby. In the Scheldt estuary in northern Europe, mollusc diversity is also higher in the freshwater marshes than in nearby saltwater and brackish communities. Reptiles found in freshwater tidal marshes in North America include the common snapping turtle, the venomous water moccasin, and the American alligator. Mammals include the marsh rice rat, Virginia opossum, marsh rabbit, mink, muskrat, river otter, and raccoon. The nutria, looking like a large muskrat, is native to South America but has been introduced into North America and Europe where it has done considerable damage to wetland plant communities. The ubiquitous white-tailed deer also inhabits tidal freshwater marshes in North America. In contrast to salt marshes, no vertebrates are restricted to freshwater tidal marshes.

Fish communities are dominated by freshwater species from the region; some anadromous fishes and marine-estuarine species use tidal marshes either as nurseries or as spawning areas. Five groups of fishes use tidal freshwater marshes in North America: obligate freshwater fishes, such as the cyprinids, centrarchids, and ictalurids; estuarine species; estuarine-marine species; catadromous species (for example, the American eel); and anadromous and semianadromous species. In these latter categories are the anadromous white shad, blueback herring, and white perch.

Tidal freshwater marshes occur in almost every coastal region of the world, yet they are relatively unstudied. They historically have been subject to conversion to agricultural uses and are threatened by the range of human impacts and activities that threaten wetlands worldwide, as well as the specter of rising sea levels.

Nontidal Freshwater Marshes

Nontidal freshwater marshes are found in nearly every geographic setting. Some of the world's most important wetlands are either predominately freshwater marshes or at least contain significant proportions of marsh: the Everglades, the Prairie Potholes of North America, the Pantanal of South America, the Sudd in the upper reaches of the Nile.

There is tremendous variation in these marshes, which compose perhaps 20 percent of the world's wetland acres. They include wet meadows, wet prairie, prairie potholes, vernal pools, and playa lakes. Freshwater marshes are characterized by grasses, sedges, and other herbaceous (nonwoody) plants and by some commonalities in their ecology.

Types and origins of nontidal freshwater marshes. The placement on the landscape of freshwater marshes is varied and reflects varied origins. All that is required for a freshwater marsh to form is a shallow depression capable of holding water for a sufficient time for hydrophytic vegetation to germinate and grow. Such depressions have many beginnings.

In parts of the world subjected to glaciation during past Ice Ages, the movement of tremendously thick, heavy sheets of ice resulted in landscapes with gently undulating surfaces, in some cases not unlike the "washboard" surfaces that form on dirt roads, except on a much larger scale. Such landscapes include features called drumlins and moraines, which are basically hills made up of rock and gravel moved by glaciers; kettles and potholes, which are small and relatively shallow; and numerous depressions in the glacial till. Such glacial landscapes are found across large areas of the northern United States and into Canada, as well as in the Russian steppe. Over time, some of these depressions have gradually filled in to the point at which they are no longer lakes but marshes, or they may be seasonal lakes that become marshes as they dry out.

Freshwater marshes also form in mountain valleys, where in some cases valley lakes have filled over the ages with eroded material from upslope as well as decomposing aquatic vegetation. In such a sequence of events, the marsh may be viewed as a transitional stage, but on a timescale of perhaps thousands of years. As the marsh fills and the ground surface rises relative to water level, it may eventually become more of a wet meadow.

Marshes also form in arid regions, such as the Great Basin of the western United States, where mountain streams flow into dry basins and form lake-marsh systems that are lakes during the snowmelt season and marshes later in the year. Some dry completely during the dry season. In others, such as Lake Chad in Africa (often described not as a lake but as a large wetland), the swings between drying out and filling up are irregular and take place on a timescale of decades to centuries, driven by climatic variations.

In permanent, stable lakes outside of the arid regions of the world, it is not unusual to find freshwater marshes around the lake margins. Sometimes the marshes are associated with deltas formed where tributary streams empty into the lake, depositing their sediments. In other cases, movement of sediments along lake shores creates barrier beaches, elevated sand formations that can cut off a shallow area from the main body of the lake, as happened at Delta Marsh, adjacent to Lake Manitoba, Canada. At nearly 54,000 ac (22,000 ha), Delta Marsh is one of the largest freshwater marshes in North America. It is shallow and nutrient rich, and supports large populations of phytoplankton and aquatic plants, such as fennel pondweed, common water milfoil, and common hornwort. In shallower areas and at the margins of open water, dominant emergent vegetation includes bulrushes, cattails, and giant reed. Wet meadows supporting grasses (for example, rivergrass), sedges (for example, wheat sedge), and sandbar willows are found at slightly higher elevations and usually are inundated only in the spring.

Delta Marsh is used seasonally by fishes and is highly suitable habitat for spawning. In winter, it is inhospitable to fishes because of freezing, but in summer, numerous species inhabit the marsh, including the fathead minnow, five-spined stickleback, nine-spined stickleback, yellow perch, spot-tailed shiner, white sucker, carp, brown and black bullheads, and Iowa darter. Like many prairie marshes, Delta Marsh is important seasonal habitat to a wide variety of birds, mostly migratory songbirds and waterfowl. During spring and fall, about 80 species of songbirds are typically identified. Flycatchers, warblers (especially Yellow Warblers, Yellow-rumped Warblers, Yellowthroats, and American Redstarts), swallows, and waterthrushes are numerous. Delta Marsh is probably best known, however, for its large numbers of waterfowl. Important species include Canada Goose, American Wigeon, Cinnamon Teal, Canvasback, Green-winged teal, Lesser Scaup, Gadwall, Blue-winged teal, Bufflehead, American Black Duck, Mallard, Northern Shoveler, Common Goldeneye, Wood Duck, Ring-necked Duck, and Snow Goose.

Freshwater marshes can also be associated with river systems. When rivers flow through broad, flat floodplains, freshwater marshes, and other wetland types are likely to be found. Some of the world's largest and most famous wetlands are of this type: the Sudd in Africa, the wetland system of the Tigris and Euphrates in Iraq, the Pantanal in South America, and the Everglades of Florida.

Where the main channel of a river meanders, oxbow lakes can form. Such floodplain lakes receive plentiful inputs of sediment and nutrients each time the river floods. Over time, they fill in with sediment and organic detritus, becoming shallow enough to be marshes and no longer lakes.

The biota themselves play an important part in the formation of marshes (as well as other types of wetlands). Wetland plants, by their structure and biomass, build organic soils, seal up leaky basins, slow currents whether riverine or tidal, and buffer the destructive energy of wind and waves. Beavers are also major builders of wetlands. Their impoundments significantly alter the hydrology of watersheds, increasing groundwater recharge and reducing runoff.

Characteristics of nontidal freshwater marshes. Freshwater marshes are based on mineral soils (see above), often with an overlying layer of decomposing plant material. While decomposition rates are relatively high, the tremendous productivity of these ecosystems supplies plenty of biomass for decomposers to work with. Soil pH tends toward neutral. Soils have nutrient levels that are higher than those of bogs, but generally lower than those of freshwater tidal wetlands. The greater the relative contribution of precipitation, the lower the nutrient levels. Table 3.3 presents some of the specific characteristics of different types of nontidal freshwater marshes.

Swamps

Swamps are what many people think of when confronted with the word "wetland," and it is from swamps that wetlands get most of their negative connotations in the popular imagination: a dark, dripping environment filled with slimy, nasty creatures and dangers (quicksand! snakes! bugs!). While swamps do resemble the popular image (they are buggy in warm weather, there are snakes, some are heavily shaded, and they are wet), they are fascinating environments with a subtle beauty that rewards those who take the time to visit them.

The distinguishing feature of swamps is the dominating presence of woody vegetation. Swamps form in many environments; their common elements are mineral soils and (like all wetlands) periodic inundation. Flooding may be deep or shallow; it may be erratic and unpredictable or regular (seasonal, tidal). Soils may be nutrient rich or nutrient poor. Some forested wetlands are almost never inundated but have saturated soils. One important type of swamp, the mangrove swamp, is associated with saltwater environments.

Types of swamps. Swamps, like wetlands in general, can be categorized in different ways: by ecoregion, by dominant vegetation, and by landscape and geomorphic characteristics. Individual swamps have a unique mix of features that are determined by all of these factors plus history and disturbance regimes. A classification based on dominant vegetation, however, can be useful (see Table 3.4).

Swamps along rivers and in alluvial landscapes (river valleys, floodplains, deltas, and other landforms created by rivers depositing alluvial materials—sand, silt, and gravel) occur in nearly every part of the world and have a wide variety of environmental conditions and plant associations. In low-gradient reaches, rivers

Table 3.3 Freshwater Marsh Types and Characteristics

Freshwater Marsh Type	Dominant Vegetation	Typical Soil Characteristics	Typical Hydroperiod	Typical Nutrient Status	Landscape Setting, Origin	Fauna	Miscellaneous
Vernal pool	Herbaceous or woody	Mineral	Seasonal (spring); relatively short period of inundation	Varies	Often small, hydrologically isolated; depressional areas; especially in mediterranean climate regions	Fish absent; large populations of insect larvae and crustaceans common; important habitat for amphibians and certain birds	May be saline in some regions; heavy losses in some regions (Central Valley of California); little regulatory protection
Wet meadow	Herbaceous (grasses, sedges); mix of wetland and upland types; high species diversity	Mineral	Seasonal; short period of inundation	Nutrient rich	Grasslands: prairie, steppe, and pampas; sometimes in floodplains, alpine valley bottoms	Important bird habitat, particularly for migrating birds	Historically used heavily by human societies; recent intensification of use including draining and elimination of native vegetation, e.g., in Europe
Prairie pothole	Herbaceous; often cyclical succession of aquatic	Mineral; glacial till	Seasonal cycle of expansion and contraction	Nutrient rich	Form in depressions of glacial landscapes of	Fish-free except where introduced; numerous	Cycle of inundation and open water, then gradual drying and

	Vegetation	Soil	Hydrology	Water chemistry	Distribution/Origin	Fauna/Habitat	Threats
	species, grasses, sedges, mudflat annuals)		superimposed on 15–20 year wet-dry cycle; often connected hydrologically through groundwater		the upper U.S. Great Plains into Canada	invertebrates; critically important habitat for waterfowl, migratory wading birds, and shorebirds	takeover by emergent vegetation, followed by complete drying out in some potholes in dry part of cycle; some potholes saline; much reduced in area and number by conversion to agriculture
Playa lake	Herbaceous; similar to wet meadow	Mineral	Seasonal; hydrologically isolated; multiple wet-dry cycles in a year not unusual	Nutrient rich due to adjacent farming	Southern to central U.S. Great Plains; apparently solutional in origin	Copious invertebrates, amphibians; important waterfowl, wading bird, and shorebird habitat	Threatened by agricultural pesticides and fertilizers and by grazing

Table 3.4 Swamp Types and Characteristics

Swamp Type	Dominant Vegetation	Typical Soil Characteristics	Typical Hydroperiod	Typical Nutrient Status	Landscape Setting, Origin	Fauna
Cypress swamp	Bald or pond cypress associated with water tupelo, black gum	Mineral	Varies, but generally inundated year-round (deepwater swamp)	Typically nutrient rich	Eastern and southern U.S. coastal plain, lower Mississippi floodplain; occurs as cypress dome, cypress strand, bottomland swamp, river or lake edge, cypress prairie	Variety of invertebrates, reptiles and amphibians, birds, and mammals as well as fishes
White cedar swamp	White cedar, sometimes with red maple and other tree species; in other cases associated with sphagnum	Mineral to organic, acidic soils	Seasonally inundated	Nutrient poor	Atlantic coastal plain of North America south to Florida; formerly widespread, now mostly isolated patches	Moose in far northern examples; variety of invertebrates, birds, reptiles, and amphibians, and mammals throughout its range
Red maple swamp	Red maple mixed with other hardwoods	Mineral	Seasonally Inundated	Variable	Perched surface depressions, low places along streams and lakes, groundwater-fed depressions	Variety of invertebrates, reptiles and amphibians, birds, and mammals; species mix depends on regional setting
Scrubshrub	Woody vegetation less than 20 ft (6 m) tall; dominant vegetation may be broad- or needle-leaved, deciduous or evergreen	Mineral, sandy, sometimes with layer of well-decomposed organic material	Lengthy, typically with little fluctuation	Typically nutrient poor, leading to dwarfism	Widespread; surface depressions; some examples represent early successional stage occurring after disturbance of forested swamp	Varies according to regional setting, dominant vegetation

deposit their sediment loads and their channels meander, resulting in characteristic topographic variations as well as variations in soil composition. Levees—linear features that are relatively high compared with the rest of the floodplain—form when rising waters spill out of the active channel onto the floodplain. The loss of turbulence as the water spreads out over the floodplain causes sediments to drop out of suspension. Levees often support forest species that are different from those in swamps on the adjacent floodplain.

In the topographic depressions behind the levees, floodwaters may persist for long periods, and tree species more tolerant of saturated, anaerobic soils dominate. Abandoned river channels form sloughs or oxbow lakes, and large areas may be subject to shallow inundation, with the levees of former channels forming low ridges with different plant associations on them. The bottomland swamp forest complexes may be quite large, as was the alluvial plain forest of the lower Mississippi River—22 million ac (8.9 million ha) or more than 34,000 mi^2 (about 90,000 km^2). The dominant trees are determined primarily by elevation, which determines hydroperiod.

Along the southeastern coast of North America, for example, deepwater swamps, characterized by long hydroperiods, are often dominated by cypress (bald cypress and pond cypress) and are thus called cypress swamps. Cypress is often codominant with water tupelo or black gum. Swamps with less lengthy periods of inundation, higher elevation, and more upland are often dominated by white cedar or red maple.

The swamp forest along the Congo River in Sub-Saharan Africa is dominated by bubinga, kratom, and a number of trees (see the Chapter 3 appendix) that do not have common names. One of these is *Alstonia congensis*, one of a number of trees that form buttress roots. Some permanently flooded areas are host to large stands of Raphia palm.

Because they are periodically subject to inundation by floodwaters, riverine swamps are less likely to be low in nutrients than swamps fed only by rainwater. However, there is significant variation in nutrient status. The *igapo* forests, growing on sandy soils of the nutrient-poor Rio Negro and Rio Xingu regions of the Amazon basin, support relatively fewer tree species and less biomass than the *varzea* forests of the more nutrient-rich Andean tributaries of the Amazon. White cedar swamps in northern North America are often nutrient poor. But these are exceptions, and riverine swamps are generally fecund places supporting a relatively high level of biomass production.

The animal life supported by riverine forested wetlands tends to be diverse and plentiful, consistent with a relatively high level of biomass production. Deepwater swamps house fishes as well as reptiles and amphibians, birds, and mammals. During periodic inundation, river fishes also invade the swamp to feed and breed, as the FPC indicates (see Chapter 2). Fishes leave floodplain and deltaic areas with shorter hydroperiods as the water recedes or become concentrated in ever-shrinking pools, as is the case in the Pantanal and the Amazonian floodplains.

Riverine swamps and bottomland forests have been affected drastically by channelization and flow regulation in many river systems. Channelization cuts off a floodplain from its river so that the supply of sediment is removed; dams may prevent flood damage but also prevent periodic inundation of floodplains, altering ecological relationships in floodplain forests.

Riverine swamps and bottomland forests have been subject to clearing, destructive logging, and habitat fragmentation in many regions. The lowland forests of the Irrawaddy delta in Myanmar are a prime example. Sediment pollution, resource extraction, and agricultural expansion have resulted in a drastic reduction in wildlife numbers and diversity. Nearly all large mammals, which included Asian elephants, tigers, and leopards, have been extirpated from this ecosystem. Bird and reptile populations are also declining rapidly, and the future of wildlife in this region is bleak.

Peatlands

Peatlands are defined as wetlands in which the rate of accumulation of organic matter is greater than the rate of decomposition and mineralization of organic matter, and where at least 1 ft (0.3 m) of peat has accumulated. Peat is the partially decomposed remains of plant material in which many plant parts—including stems and leaves—are still identifiable. Peatlands cover more than 400 million ha (almost 1,000 million ac), or about 3 percent of the world's land area. Known as mires in Europe where they are common, peatlands make up more than half of all wetlands in the world. In the high latitudes of the Northern Hemisphere, they stretch for hundreds or even thousands of square miles. But peat-forming wetlands are not limited to the high latitudes. In the extensive peat forests of the Tropics the organic material that accumulates is wood. Northern peatlands consist of two types of wetlands: bogs and fens.

Bogs are wetlands with soft, spongy, organic soils, in which the dominant plants are usually sphagnum moss, shrubby ericaceous plants, and conifer trees, usually at the margins. These plants form a dense mat that may actually float detached from underlying peat layers. Bogs and fens with such floating mats are sometimes called quaking bogs. Peat accumulates beneath the layer of living plants because the annual addition is greater than the annual rate of decomposition. Decomposition is slow, primarily because of acidic, anaerobic conditions, and also because dead sphagnum moss resists decomposition. These conditions are not only suitable for sphagnum, but to some degree are created by sphagnum. Sphagnum has been termed an ecosystem engineer, because it modifies its environment. Once it enters a wetland and becomes established, it holds water and modifies water chemistry, making it more acidic and removing nutrients. On the other hand, it has been found in Europe that Scots pine, after germinating in a sphagnum-dominated wetland, alters the bog environment in a way that is fatal to the sphagnum but beneficial to the maturing pine.

Bogs ("true" bogs or ombrotrophic bogs) receive all or virtually all of their water from precipitation, and therefore the plant and animal community that inhabits them could be described as nutrient and mineral deprived. Their low-nutrient status gives rise to a number of interesting adaptations, the most famous of which is probably the development of carnivory in plants. Plants such as the pitcher plant, sundew, and venus flytrap, all bog plants, trap and digest insects to supplement their meager intake of essential nutrients (see Plate VI).

Fens are in some respects intermediate between swamps and bogs. Like bogs, they are peat-forming, so that their soils are organic, not mineral; unlike ombrotrophic bogs, they are open systems that receive water from surface- and groundwater sources. Because of the origin and flow of water through them, there is greater availability of nutrients and minerals than in a bog. Fens are less acidic than bogs and may even be slightly alkaline. The through-flow of water prevents or reduces the buildup of byproducts of plant life and decomposition that create the harshly acidic environment of ombrotrophic bogs. Fens generally have greater plant and animal species diversity than bogs, in part because they have little or no sphagnum. Rather, they are dominated by sedges such as cattail and grasses.

Fens often form in surface depressions that receive surface- and groundwater from a surrounding watershed. But through the centuries, the accumulation of peat raises the level of the fen to the point at which it no longer receives surface- or groundwater inputs, at which point it becomes a bog. This is how some ombrotrophic bogs are formed, and bogs formed this way are termed ombrogenous bogs, or sometimes raised bogs. The spread of bogs beyond the boundaries of the original basin or surface depression, essentially taking over the (formerly dry) landscape, is referred to as paludification. Blanket bogs (see below) have their origins in the process of paludification.

Another common way for bogs (and fens, depending on the water source) to form is through transformation of a lake. A floating mat of sphagnum, or in the case of a fen of sedges and sphagnum, gradually grows outward from the lakeshore toward the center. Eventually, peat and sediment build up underneath the mat, and the mat grows to the point at which there is no more open water. This process is referred to as terrestrialization.

Fens form in low areas in the landscape and never rise above the surrounding land, so they continue to receive groundwater, if not surface water, inputs. Sometimes called topogenous bogs, these fens are fairly common in glaciated landscapes, where surface depressions are abundant. One type of topogenous bog is known as a blanket bog. This is a true ombrogenous bog that simply covers a relatively flat-lying landscape like a blanket. Conditions necessary for the formation of a blanket bog include abundant rainfall and a cool climate.

Limnogenous bogs and fens (also known regionally as pond border bogs) form on the edges of surface water bodies—lakes and low-gradient rivers. Soligenous bogs and fens develop in depressions on slopes where groundwater trickles or seeps out.

Tropical peatlands are estimated to cover 72 million ac (29 million ha) worldwide. Much of it lies in the Indomalayan biogeographic region (Southern Asia,

Southeastern Asia, Indonesia, and Malaysia), where peat swamp forests form behind mangrove forests. Dominant vegetation is a mix of tropical tree species, including ramin, jongkong, alan, and sepetir. The peat consists primarily of the partially decomposed woody roots, trunks, and branches of this vegetation. Diversity of animal life is great, with many species from adjacent lowland rain forests also occurring in the peat swamp forests. Macaques, monkeys, gibbons, and orangutans are found along with a profusion of tropical birds, bats, and insects. Peat swamp forests of Indomalaysia are under tremendous pressure from logging and conversion to agriculture. Draining of these swamps has led to recent catastrophic peat fires that released massive amounts of carbon dioxide into the atmosphere.

Regional Examples of the Wetlands Biome

Descriptions of four large wetland complexes follow: two in mid-latitude settings, one in a low-latitude setting, and one in a high-latitude setting.

Mid-latitude Wetlands: Prairie Potholes and Playa Lakes of the North American Great Plains

Prairie potholes. Prairie potholes range from wet meadow to permanent marsh. Although similar wetlands occur in other formerly glaciated grasslands (for example, in Eurasia), prairie potholes are restricted to the Prairie Pothole region of North America. This region of almost 300,000 mi^2 (almost 800,000 km^2) includes parts of North and South Dakota and Minnesota in the United States, and Manitoba, Saskatchewan, and Alberta provinces in Canada (see Figure 3.4). The prairie potholes were formed by the advance and retreat of Pleistocene glaciers across the landscape. The climate of the Prairie Pothole region is continental, with hot summers and cold, relatively dry winters. Annual precipitation is variable but generally semiarid. Although most pothole wetlands are freshwater systems, some have relatively high salinity levels due to high evapotranspiration rates in this generally dry area.

These mostly small, unconnected wetlands originally covered an estimated 30,000 mi^2 (78,000 km^2), but more than half have been lost to agricultural use since Europeans settled the Great Plains. Still, 3.1 million wetland potholes remain in the United States, and the Canadian Prairie Pothole region is estimated to have between 4 and 10 million potholes. Three-quarters are less than 1 ac (0.4 ha) in size, but the overall range is from much less than an acre to several square miles. Potholes tend to be shallow (less than 5 ft/1.52 m) and saucer-shaped. They only lose water by evaporation, except during exceptionally wet periods, when water spills over the saucer's edge.

The hydroperiod of prairie potholes responds both to seasonal and longer term (10- to 15-year) wet-dry cycles. Wet years and dry years are both necessary to keep the pothole wetlands from turning into ponds, on the one hand, or losing their

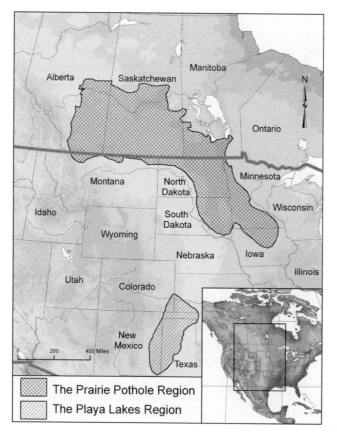

Figure 3.4 Prairie Pothole and Playa Lakes Regions of North America. *(Map by Bernd Kuennecke.)*

wetland character altogether on the other. The dramatic variability in precipitation from year to year results in changes in both the size of individual potholes and in the number of pothole wetlands that can be counted in any particular year.

The wet-dry cycle causes a cycle of changes in the wetland character of prairie potholes. In deeper potholes, dry marshes (with no standing water) result from several dry years during the cycle. Increasing precipitation creates dense regrowth of wetland plants. Aquatic and semiaquatic emergent vegetation is able to reemerge after the dry years because seeds in the sediments remain viable. As water levels continue to rise during the wet years, central parts of the pothole become too deep to support emergent vegetation, and open water appears in the middle. Further increases in water levels bring a predominance of open water, with emergents surviving only around the margins. The endpoint of the wet cycle finds open water predominating and most emergents eliminated. In potholes too shallow for water to reach a depth sufficient to eliminate emergent vegetation, the plant species occupying the wetland change.

At different stages of the wet-dry cycle different plant groups dominate: wet prairie perennials, sedge meadow perennials, shallow emergent perennials, deep emergent perennials, submerged aquatics, floating annuals, mudflat annuals, and (rarely) woody plants. Animals of the prairie potholes respond to the wet-dry cycle, too. Numbers of herbivores and insectivores (water-associated birds, for example) rise and fall with the number of potholes filled with water, and animals that prey on them (for example, foxes, coyotes) experience dramatic population booms and busts as well.

Vegetation zonation with respect to distance above or below the water level is also evident at any point in the cycle. Zonation means that there are distinctive plant associations made up of plants with similar requirements for germination and survival with respect to saturation and soil chemistry. Larger potholes have more zones than smaller potholes. Zones with longer periods of inundation throughout the year have taller emergent vegetation but fewer different species than those that are drier. Plants characteristic of the higher elevations, least often inundated, include goldenrod, Kentucky bluegrass (introduced), giant sandreed, and smooth brome. At about 10 in (0.25 m) lower elevation, Timothy and common spikerush begin to appear. At about 2–2.5 ft (0.6–0.8 m) lower elevation compared with the highest zone, reedgrass and upright sedge are found, but goldenrod and giant sandreed are not. Moving lower still, to a zone with a considerably longer inundation period, one may find, in addition to reedgrass, wheat sedge, blister sedge, water knotweed, broadfruit burreed, and cattail.

Salinity of prairie potholes ranges from freshwater levels to highly saline. In moderately saline potholes of Eastern Montana, hardstem bulrush is the dominant emergent species. Alkali bulrush predominates in higher salinity wetlands, but when salinity levels are very high, emergent plants are generally absent and potholes take on the character of mud flats with greasewood, saltbush, or pickleweed growing sporadically.

Most biomass produced by the plant community (and productivity is generally high) is not consumed directly, but rather dies and becomes available as food energy only after the detritivore community (primarily invertebrates) has processed it. Recent studies suggest, however, that invertebrates in these systems rely heavily on consumption of algae as well as decaying macrophytes. Invertebrate populations support populations of larger, predatory invertebrates as well as a host of amphibians and birds.

It is the birds for which the prairie potholes are known and for which they are conserved. Millions of birds, waterfowl, and shorebirds (see the Chapter 3 appendix) arrive to breed in the spring and remain throughout the growing season. Roughly half of the ducks included in the most popular game species originate in the Prairie Potholes region. Passerines (perching birds), hawks, vultures, falcons, and other birds not directly associated with water also nest or shelter in prairie potholes.

As the prairie potholes cycle from dry marsh to open water to dry marsh, different habitat conditions make them attractive for different bird species. The large numbers of ground-nesting birds attract many predators, including foxes, mink, raccoons, skunks, weasels, and badgers. The potholes are also home to a number of reptiles and amphibians, but their unconnectedness and tendency to dry up periodically make them unsuitable for fishes. Fishes such as the fathead minnow have been introduced in some potholes for the bait fish industry.

Playa lakes. Playa lakes are superficially similar to prairie potholes but are different on several counts. First, they are located in a different region (see Figure 3.4). Although it has been reported that playa lakes are found in a number of areas around the world, by far their greatest concentration is in the high plains of the United States, an area that includes at least the Oklahoma Panhandle, the Central Texas Panhandle, the southwestern corner of Kansas, and the eastern edge of New Mexico. Here, there are between 25,000 and 30,000 playa lakes, with the majority in Texas. However, recent studies taking advantage of digital aerial photography, satellite imagery, and soils data have included a large area of western Kansas and far western Nebraska not previously described as playa country. If these new areas are included, the count goes to more than 62,000 playa lakes. Playa lakes are (confusingly) to be distinguished from playas, which are the remnants of dried up endorheic lakes. Although playas may contain shallow lakes in the wet season, most of the time they are dried salt beds.

Another difference between playa lakes and prairie potholes is in the process by which they are formed. It is thought that on the high prairie, when a small depression would begin to collect water, acidic conditions from decomposing organic material would dissolve the underlying carbonate material, called caliche. The loss of this material through leaching downward into groundwater would cause an expanding area of surface subsidence, resulting in a growing, flat-bottomed depression shaped much like a pie plate. Depth is uniform except at the edge and is rarely more than 3.3 ft (1 m).

Playa lakes receive their water from precipitation in their respective watersheds. Precipitation in the playa region of the Great Plains averages 13 to 18 in (33 to 45 cm) annually, but actual precipitation is usually not average but either far below or far above. Precipitation occurs in late spring and early fall.

Each playa lake is the collecting basin for surface runoff and precipitation in its watershed, and groundwater usually makes no contribution. However, playa lakes do serve as recharge areas for groundwater, so not all water loss is due to evapotranspiration. Watersheds are small, averaging 137 ac (55.5 ha), but then the playa lakes themselves are small, with an average basin area of 15.5 ac (6.3 ha). Although studies of playa hydrology are few, limited data suggest that most playa lakes are inundated at least once every two years, and when flooded, they have water in them for at least two weeks during the growing season. The duration of ponding was between 1 and 32 weeks, but the latter figure was in an exceptionally wet year.

Zonation of vegetation is seen in playa lakes as it is in prairie potholes. Vegetation patterns even in the same playa can change dramatically from year to year because of variable precipitation and because of cultivation during dry years. The plants of playa lakes with short hydroperiods resemble nearby upland vegetation and include such perennial species as western wheatgrass, buffalograss, and vine mesquite. Playa lakes with longer hydroperiods have more characteristic emergent wetland vegetation. The most commonly encountered playa lake plant associations are wet meadow type, dominated by barnyard grass, pale spikerush, and mucronate spangletop; and broadleaved emergent type, dominated by various species of smartweed.

Playa lakes support a wide variety of animal life and are particularly important habitat in the dry southern and central Great Plains. For example, the playa lakes of the southern Great Plains provide wintering habitat for an estimated 2 million waterfowl of about 20 species. The lakes' abundant populations of macroinvertebrates, particularly insects, make them an important rest and "refueling" stop during the spring and fall migrations of millions of birds. During the initial inundation of playa lakes, crustaceans such as seed shrimp and fairy shrimp predominate; later these are eclipsed by increasing numbers of insect larvae. In general, invertebrates populations and communities are highly dynamic and respond to short-term changes in hydrology in these ephemeral water bodies.

In one study, 30 species of shorebirds were found to be using these lakes. The species with the greatest numbers during the spring migration were American Avocet, Long-billed Dowitcher, and Wilson's Phalarope. During the fall migration, the most abundant species were American Avocet, Long-billed Dowitcher, Long-billed Curlew, Stilt Sandpiper, and Lesser Yellowlegs. The Long-billed Curlew and the Snowy Plover nest in the region.

For the most part, playa lakes in the Great Plains provide the only habitat in that region for animal species that need wetlands. Amphibians use playa lakes and probably would not exist in the region without them. They also support a number of reptiles and mammals.

A Low-Latitude Wetland: The Pantanal

The Neotropical biogeographic realm (South and Central America) contains some spectacularly large wetland systems, some of which are primarily forest. Numerous wet grasslands and associated freshwater marshes, some very large and exceptionally diverse in both plant and animal species, also occur. Among them are the *llanos* (flat grassland) of Colombia and Venezuela, the *llanos de moxos* of Bolivia, the flooded savanna of the Parana River, and the Pantanal.

The largest wetland system in the Neotropical region, and one of the least affected by human activity, is the Pantanal, rich with plant and animal life, varied in its habitats, and ever changing. As with many large wetland systems, it is difficult to categorize, in part because it contains such a diversity of environments, and in part because its sheer size puts it in a category by itself. Imagine a wetland of 54,000 mi^2 (140,000 km^2). This is the low end of a range of estimates; some researchers put the area at 81,000 mi^2 (210,000 km^2). Even the low estimate would

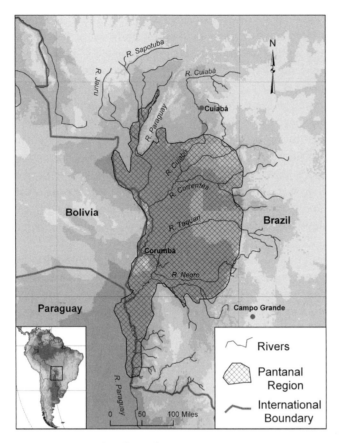

Figure 3.5 The Pantanal and its tributaries. *(Map by Bernd Kuennecke.)*

make the Pantanal about the same size as the state of New York, and larger than Pennsylvania or England.

The Pantanal is in the upper watershed of the Paraguay River, a very large tributary of the Parana River (see Figure 3.5). The watershed of the Paraguay/Parana river system abuts that of the Amazon, draining land to the south of that great river basin. It empties into the Atlantic at Buenos Aires, Argentina. About 80 percent of the Pantanal is in Brazil, in northern Mato Grosso and the southern Mato Grosso do Sul states. A fifth of the Pantanal lies in Bolivia and Paraguay, west of the Paraguay River.

Surrounded by highlands that include the Gran Chaco of Bolivia and Paraguay and the Brazilian *Planalto* (highlands), the Pantanal is low lying (250–660 ft or 75–200 m above sea level) and flat, like an enormous flat-bottomed bowl. The basin is old (65 million years), and sediments from the Pantanal's tributaries, particularly the Taquari River, have been accumulating in alluvial fans for millennia. During past glacial periods, a saltwater lake formed in the basin.

Climate and hydrology. Precipitation is strongly seasonal. During the rainy season, rain pours down on the Pantanal and runoff from the highlands flows into it. Annual average rainfall is not unusually great, about 32–48 in (80–120 cm), but it is concentrated in the October-March rainy season. In the uplands north and northeast of the Pantanal, average annual rainfall is closer to 48 in (120 cm) and even more concentrated in time. However, long term, the climate is quite variable, subject to multiyear wet and dry periods. Prehistorically, the Pantanal suffered from severe drought during glacial periods, and the comparatively low number of endemic species (less than 5 percent for most taxa) has been attributed to this fact.

Annual floods leave only small islands and natural levees of dry ground scattered about. The water flow is from north to south and east to west and is slow, because the land slopes imperceptibly from north to south and from east to west. Because the slope is so low, the period of flooding lasts for months.

Plants. While the regional climate has two seasons, wet and dry, four seasons of life are recognized in the Pantanal, corresponding to the waxing and waning of the floods. The *Enchente* begins at the start of the rainy season (around November into the austral summer, although it is variable) and is the season of rising waters. The *Cheia* begins after flooding reaches its maximum, depths of about 16 ft (5 m) in some parts of the floodplain. Temperatures are cooler, and nearly all of the Pantanal is underwater. The floodwaters, which have some degree of turbidity from sediments, now become extremely clear, making possible a luxurious growth of underwater plants in addition to emergents in shallower water. Next, the *Vazante* comes, with its rapidly falling water levels, and aquatic plants are exposed and die, to be replaced by terrestrial species. Finally comes the cool dry season, the *Seca*, in which large parts of the Pantanal dry out. Except for the deep channels of the larger rivers, the only water left is in shrinking shallow lakes. As they shrink, massive die-offs of plants and aquatic animals, low-oxygen conditions, and algal blooms begin. A concentrated food supply attracts huge numbers of birds and other wildlife. At the end of the *Seca,* air temperatures rise and water temperatures reach their annual maximum. Then the rains begin again. It is an annual cycle of extremes of wet and dry conditions to which plants and animals have adapted over the millennia.

Because the Pantanal's enormous size gives rise to heterogeneity in environmental conditions, scientists who study it find it useful to distinguish different regions. The northern Pantanal differs from the southern Pantanal in climate and hydrology. The south is cooler, and its floods peak months later than those of the north. Maximum flood levels are reached as early as February in the north and as late as June in the south. The rivers that feed the Pantanal, because of the geographic positions of their basins, tend to flood sequentially so that the hydroperiod—the period of flooding—is extended. The sequence is rarely the same from one year to the next.

As with wetlands everywhere, hydroperiod and elevation are strongly correlated. Thus, the Pantanal is a mosaic of habitats distinguished by small differences in elevation, which result in large differences in vegetation. Furthermore, the plant communities in areas that undergo a flooding-drying cycle demonstrate pronounced succession over the course of the year. Ecologists have distinguished as many as 16 major plant associations in the Pantanal, but these can be reduced to three broad types: natural levees and ridges that are never flooded and support gallery forests; seasonally flooded grasslands and wet meadows; and surface water bodies with aquatic and emergent vegetation.

At the lowest elevation is the network of rivers and streams, some of which become intermittent strings of lakes during the low water period. At about the same level are lakes and ponds ranging from a few feet in diameter to many miles. These may be permanent or in existence only during the wetter months. One of the Pantanal's characteristic plant community types, floating mats, inhabits these lakes but may float away during the floods or dry up and burn during the dry season. These floating marshes, by some estimates, contribute more biomass than any other vegetation in the Pantanal. Their plants include water hyacinth, eared watermoss, Cuban bulrush, horsetail paspalum, and water lettuce; and they support huge numbers of algae, microcrustaceans, insects, and fishes. The lakes and littoral zones where these floating marshes appear also support a diverse mix of underwater and emergent vegetation. Emergents include giant flatsedge, softstem bulrush, and southern cattail.

Seasonally flooded grasslands (*campos*) and savannas cover about 70 percent of the Pantanal. These may be inundated by river flooding or by rainwater; in the dry season, they are subject to fires. Habitat types found in the grasslands are typical of the Brazilian cerrado and include park savanna, palm savanna, wet meadows, and marshes. Dominant species include carpetgrass, spikerush, and panicgrass, with caronal grass at slightly higher elevations. Scattered palm trees and scrubshrubs dot the landscape. The savannas grade into semideciduous alluvial forests with small trees and bushes. Areas flooded by rivers rather than rain have more clayey soils and support park savanna dominated by trumpet-tree as well as palm savanna. There are also extensive "termite savannas" with termite mounds a prominent feature of the landscape, along with various tree and shrub species including annona and blackrodwood.

On the rarely or never-flooded areas of higher ground and on the natural levees both active and abandoned, there may be semideciduous, deciduous, and gallery forests (in some cases, predominately of palm trees). Important tree species of this habitat include cohoba, cottontree, guajava, acacia, kapok tree, earpod tree, mimosa, piptadenia, and guayacan.

Animal life of the pantanal. While the biodiversity of the Pantanal may not equal that of its neighbor to the north, the Amazon basin, it is home (by one conservative account) to 498 species of lepidopterans (butterflies, moths, and skippers), 264 species of fish, 652 of birds, 102 of mammals, 177 of reptiles, and 40 of amphibians. The fauna of the Pantanal (see Plate VII) is largely derived from the cerrado, the dry Brazilian woodland-savanna ecoregion that borders the Pantanal on three sides, together with the Amazon; its rate of endemicity is not especially high.

Invertebrates are extremely abundant in the Pantanal. The most diverse and richest invertebrate fauna is probably that inhabiting the underside of the floating meadows. It is a veritable wonderland of rotifers, turbellarians, gastropods, nematodes, aquatic worms, tiny crustaceans (cladocerans, copepods, amphipods, and conchostracans), and aquatic insects both larval and adult. Among the aquatic insects, most orders are represented but particularly common orders are caddisflies (Trichoptera), true flies (Diptera), true bugs (Hemiptera), and beetles (Coleoptera).

Though many arthropods in the Pantanal require aquatic habitat at some point in their lives, the cycle of flooding and extreme aridity is challenging for most. The Lepidoptera (butterflies, moths, and skippers) are an outstandingly diverse group, but so are other insects and spiders: one study of three palm trees during the high-water period found almost 16,000 individual arthropods, of which 87 percent were insects and 13 percent were spiders. The dominant groups were Hymenoptera (ants, wasps, bees), mostly ants, followed by Coleoptera (beetles) and Araneae (true spiders). Among the 2,197 adult beetles collected were 32 families and 256 species. The mix of species and their relative numbers are affected, like everything else living on the Pantanal, by seasonality (wet versus dry). It is likely that many terrestrial insects and spiders perform a cyclical migration into and out of trees, similar to the movement of fishes into and back out of the flooded lands.

Molluscs are especially diverse in the Pantanal. The many lunged and gilled snails (Gastropoda), mussels, and clams are eaten in huge numbers by otters, wading birds, and other animals including humans. Numerous freshwater crabs and prawns (order Decapoda) are important links in the food chain between detritus and larger predators, including caimans.

The number of fish species in the Pantanal is estimated at between 268 and 405. They belong mostly to the orders Characiformes, Perciformes, and Siluriformes. Characiformes includes characins, pencilfishes, hatchetfishes, piranhas, and tetras; characins are particularly numerous in the Pantanal, but piranhas are more notorious. Of the Perciformes, the cichlids are especially well represented. The Siluriformes include several species of long-whiskered catfish (family Pimelodidae) and suckermouth armored catfish (family Loricariidae).

Fishes largely follow a cyclical pattern of activity over the course of a year. During the *Enchente*, as the rivers flood and water begins to rise in the vast plains of the Pantanal, fishes disperse into the flooded areas, some after having migrated upstream when the waters began to rise. Then, as water levels fall in the *Vazante*, they move back toward rivers or into lakes and ponds. Many are consumed by predators as they become concentrated in shrinking pools. During the intensifying *Seca*, low oxygen levels and predation continue to inflict losses on fish populations. Many live out the *Seca* in rivers, while others burrow into the mud and become dormant.

This is necessarily a vast oversimplification of an extremely complex and dynamic food web. The complexity and dynamism are due in no small part to the seasonal ecological succession driven by the hydrologic cycle of the Pantanal. Fishes of the Pantanal have adapted in myriad ways to the pressures and opportunities created by the cycle of extreme wet and dry conditions. The feeding behavior of the pacu caranha illustrates this point. During the dry season, when it inhabits mainly the river channels and lakes, this fish eats primarily invertebrates such as freshwater crabs; but when fruits and flowers are available during the periods of flooding, it switches to those. Many small characids live on the invertebrate community among the roots on the underside of the floating marshes, but they fan out into the flooded meadows to forage for food during the flooding seasons. Many invertivores—fish that feed on invertebrates—switch from one prey species to another as successive invertebrate population peaks occur. Fishes, in other words, are opportunists in the Pantanal.

Many of the Pantanal's most commercially important fishes are migratory. As water levels drop during the *Seca*, migratory adults move back into the river channels and begin swimming upstream to the headwaters. This migration, known as the *piracema*, is generally led by characid species. Arriving in the headwaters, they reproduce and wait until the rains begin and the rivers rise. At this point they release their eggs, which are swept downstream. The eggs and fry (small hatchling fish) are carried by rising floodwaters into the shallow flooded meadows and plains of the Pantanal, where they find plentiful food and shelter from predators. The adults then begin a return migration called the *rodada*. They arrive emaciated as floodwaters begin to spread over the plains, where they will spend the high-water months. As the floods recede, the cycle begins again for the adults. The young remain downstream in floodplain lakes until they reach adulthood. During the *Seca*, mortality from predation and starvation may be high.

Migratory species include many characins, of which the best known are the curimbata; tambaqui; and silurids (catfish). Some of these migratory fish species are remarkable for the distance of their migration, which may be many hundreds of miles. Their migrations have historically attracted so many subsistence and commercial fishermen that restrictions have been placed on the exploitation of some species.

A number of fishes are endemic to the Pantanal, or at least to the Paraguay River basin. These may include an entire genus of freshwater flounder (*Hypoclinemus*). A number of characins are reported to be endemic to either the Pantanal, the Rio Paraguay, or the Rio Parana/Rio Paraguay system. The majority of fishes in the Pantanal, however, are species of Amazonian origin.

The Pantanal is a birdwatcher's paradise. The numbers of birds reportedly inhabiting the vast wetland range from more than 400 to more than 800, depending largely on how the region is defined. It is not just the number of species that makes the Pantanal so remarkable, but the huge population sizes of birds that can be found there. Many birds are resident, but large numbers of migratory birds also spend time in the Pantanal.

While species diversity and abundance may be high, the rate of endemicity is low (2 percent). The Pantanal is a corridor linking neighboring biogeographic regions. Open landscape cerrado and Chaco birds and a smaller percentage of forest species move between the Amazon and the South Atlantic rainforests. Other migratory birds, including Nearctic migrants, use three flyways that converge in the Pantanal. For other birds, the Pantanal is a barrier, creating either the southern limit of Amazonian species (probably because of occasional outbreaks of cold air from the south) or the northern limit for south Atlantic coast species. The pattern is similar for lepidopterans (butterflies).

Many species and large numbers of waterbirds live on the Pantanal. Some are permanent residents, others are migratory—including a significant number of Nearctic migrants. The waterbirds include some of the largest and showiest of the Pantanal's birds, such as the jabiru stork or *tuiuiu*, a symbol of the Pantanal (see Plate VII). Scores of ducks, cormorants, herons, grebes, rails, bitterns, kingfishers, kites, and even a swallow are included among the waterbirds of the Pantanal.

As befits an environment that alternates between wet and dry, many non-waterbirds also inhabit the region. Most are species that inhabit the nearby *cerrado*. Among them are 29 species of hummingbirds (family Trochilidae) and a large flightless or nearly flightless bird, the rhea. The crested caracara, a member of the falcon family (Falconidae), is sometimes a predator but also scavenges the plentiful roadkill on the Pantanal's few roads. Another spectacular bird of the Pantanal is the very large, bright blue Hyacinth Macaw (see Plate VII), one of the largest parrots in Brazil—or the world for that matter—and critically endangered due primarily to poaching for the international trade in rare birds. A beautiful cardinal, the Yellow-billed Cardinal with its striking red head, is emblematic of the Pantanal. One endemic bird is the Mato Grosso Antbird.

Rookery trees (known as *ninhais*), festooned with hundreds if not thousands of large birds, moving and making a tremendous racket, present one of the Pantanal's more memorable impressions. Colonies of several different bird species will occupy the same ninhais, segregating into vertical zones. The excreta of so many birds is often fatal to the tree, and the concentration of nests full of eggs and chicks, some of which fall out, attracts many predators, both terrestrial and aquatic.

While some mammals are present on the Pantanal in large numbers, diversity is low; various accounts put the number at a little over 100. A third may be bats. No mammals are endemic. Some of the more salient species include the world's largest rodent, the capybara, a peaceable herbivore basically terrestrial but quite at home in the water. They are hunted for both their fur and their meat and are preyed upon by large carnivores. Deer in the Pantanal are diverse; among five recorded species is the semiaquatic swamp deer. Deer populations are declining because of competition with the increasing number of cattle grazing in the Pantanal. Peccaries (collared and white-lipped) are also plant eaters in the Pantanal.

Carnivores that prey not only on mammals but also on birds, reptiles, amphibians, and even fish, include four wild dogs and several cats. One of the wild dogs is the rare and endangered maned wolf or guara wolf of the cerrado. The largest of the cats is the jaguar, but another big cat, the puma, is more numerous. The

Pantanal jaguar has much in common with the tiger, including a penchant for swimming. Several smaller felines are declining in number: the jaguarundi, the tiger cat, and the ocelot. Aquatic carnivores include the giant otter or *ariranha* (see Plate VII) and a lesser otter.

Two monkeys are common to the Pantanal: the howler monkey and the smaller brown capuchin. Both are arboreal, and the howler monkey's guttural vocalizations can be heard for several miles; it may be the loudest animal in the New World.

The Pantanal has in the past century acquired several new mammals: the millions of domestic cattle on the ranches that cover the vast majority of the land, domestic water buffalo, feral water buffalo, feral pigs, and feral cows.

A wetland as large as the Pantanal should be a cornucopia of amphibians, and it is, at least seasonally, when amphibians abound. Nonetheless, diversity and endemicity are relatively low. In the northern Pantanal, 30 amphibians have been identified. About half of the frogs and toads are arboreal. Populations increase with the expansion of habitat that comes with the end of the dry season, but are subject to fearsome predatory pressure from carnivores of every kind: birds, reptiles, and mammals.

Two species of anaconda live in the Pantanal. The yellow anaconda can reach 20 ft (7 m) in length and hunts mainly in the water, whereas the similar-size boa constrictor is terrestrial or arboreal. The Pantanal coral snake (*Micrurus tricolor*), the Neotropical rattlesnake, and two lanceheads (the Brazilian and the Neuwid) are all poisonous. Lancehead venom is highly toxic and can kill in a matter of hours. The Neotropical rattlesnake creates little damage where the bite occurred, but may result in blindness, paralysis, and respiratory failure.

Probably the reptile most closely associated with the Pantanal, by virtue of its large numbers and visibility, is the caiman or *jacare* (see Plate VII). Three species of caiman were historically recorded in the Pantanal, but only one apparently remains, and it is doing well, being extremely numerous throughout the Pantanal. It is a major player in the food web and is sometimes credited by local inhabitants with keeping down the populations of piranhas. An opportunistic carnivore, the jacare feeds on molluscs, crabs, fishes, birds, capybaras, and other mammals, and even other jacares. It is the focus of intensive conservation efforts, as poaching for its skin is a clear threat.

The future of the Pantanal. At a recent symposium on conservation of the Pantanal, comparisons were made between the state of the Pantanal today and the state of the Florida Everglades 50 years ago. Then, the Everglades was still a relatively healthy ecosystem, and its sheer size and apparent robust ecological health made it seem invulnerable. But the threats that eventually would bring about its collapse were already evident to those familiar with the region: large-scale water development schemes that altered a critical environmental variable, hydrology; deforestation for agriculture; draining of large areas for pasture and cropland; fragmentation by roads; introduction of invasive species; pollution; poaching and overhunting; and increasing human and livestock populations. A similar suite of dangers gathers about the Pantanal today, and if they grow as many expect, the Pantanal will go the way of the Everglades, into a downward spiral of ecological degradation and collapse.

A High-Latitude Wetland: The West Siberian Lowlands

The West Siberian Lowlands (WSL) region is variously described as the world's largest lowlands and the world's largest wetland. The WSL is indeed an enormous landscape of more than 1 million m^2 (2.7 million km^2), sparsely inhabited and largely covered by marshes and peatlands (see Figure 3.6). Its western border is the Urals, a north-south running mountain chain 746 mi (1,200 km) west of Moscow. On the east, about 1,240 mi (2,000 km) from the Urals, the region is bordered by the Yenisei River, which flows north from northern Mongolia. The southern border of the lowlands proper follows roughly the 55° N parallel, from Chelyabinsk in the west, through Omsk, Novosibirsk, and Krasnoyarsk to the east. South of that line, the landscape begins a transition to the higher and drier land characteristic of the Russian steppe.

The southern half of the WSL drains into the Ob-Irtysh river system and the eastern edge into the Yenisei. The remainder discharges to several smaller rivers (though still sizeable) that flow into the Ob's estuary. The Ob-Irtysh watershed is

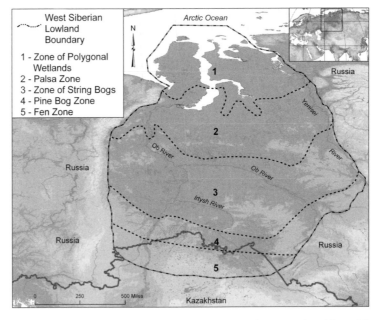

Figure 3.6 The West Siberian Lowlands. *(Map by Bernd Kuennecke. Adapted from Fraser and Keddy 2005.)*

the fourth-largest river basin in the world, comprising 1,158,302 mi^2 (nearly 3,000,000 km^2); its watershed includes most of the WSL. The basin begins in the steppes of Kazakhstan and in the mountains that form the border with China and Mongolia. Once the Irtysh arrives in the WSL, it travels 1,550 mi (2,500 km) and only drops about 500 ft (150 m). The Ob also has a leisurely trip through the low-lands due to its extremely low gradient. Most of the WSL is less than 330 ft (100 m) above sea level.

The WSL landscape is flat. It is a mosaic of differing vegetation types, with meandering rivers winding through a country dotted with lakes and ponds. Aerial photos and satellite images reveal the occasional human settlement, highway, or pipeline traversing the vast expanses, but for the most part it is sparsely inhabited. Small wonder, as the climate is inhospitable in the extreme. It is highly continental, meaning very cold winters and warm summers. The temperatures can go from as low as $-75°$ F ($-60°$ C) in the winter to 90° F (35° C) in the summer. Usually, how-ever, the summers are more moderate and wet. Mostly they are short. Depending on how far north in the WSL one is, the growing season may be as few as 50 days long.

Zones of the WSL. The predominant vegetation follows a pattern of zones with changing latitude: the Zone of Polygonal Wetlands, the Palsa Zone, the Zone of String Bogs, the Pine Bog Zone, and the Fen Zone. The Zone of Polygonal Wetlands is the northernmost zone of the WSL. Here, permafrost underlies all soils, and the vegetation is essentially that of the tundra—grasses and sedges, heaths, mosses, lichens, and some forbs (broadleafed plants). It accounts for approximately 13 per-cent of the WSL, or 138,000 mi^2 (357,000 km^2). Bogs cover the majority of the area.

Polygonal wetlands apparently originate when frost cracks associated with the formation of permafrost create polygonal shapes on the landscape. Over time, low "walls" of sediment and vegetation form along these cracks, enhancing a heteroge-neous landscape of polygons 30–100 ft (about 10–30 m) in diameter. These walls are perhaps 1 ft (0.3 m) high and 1.5 feet (0.5 m) wide. Drier, slightly elevated con-ditions on the walls favor the mixed growth of dwarf shrubs, mosses including sphagnums, and grasses and sedges. The poorly drained, sheltered low interiors of the polygons support a different mix of grasses, sedges (such as *Carex stans*, water sedge), and mosses adapted to oligotrophic conditions. Such conditions, with low nutrients and mineral availability as well as relatively low pH, are well suited for the formation of peat.

Much of the region, particularly the lowest areas (floodplains, lake edges), is covered by homogenous wetlands—flat, unrelieved expanses of peatland, some-times dotted by tussocks that rise above the monotonous low level of the other plants. Tussocks are clumps or tufts of cotton grass that are elevated by roots and older plant material.

Winters are long, cold, and windy; summers are brief and cool. Vegetation shows various adaptations to the severe conditions. First, it is all close to the ground, where there is some shelter from the drying, damaging winds. The few trees and shrubs (for example, some willow species, such as the polar willow) are dwarfed in the extreme. Second, some plants such as members of the heath family (Ericaceae), which includes azaleas and rhododendrons, have tough, leathery leaves that can conserve moisture against the drying effects of the wind. Third, some plants adopt growthforms that could be described as huddling together against the elements: growing in clumps or masses. Grasses and sedges form tussocks, and heath species also grow together to form cushions or mats. The former is not only an adaptation against the cold and wind, but also a defense against varying water levels. Fourth, some plants, forbs for example, adopt a rosette-style growth pattern in which the tender new growth is surrounded by concentric rings of living and dead leaves, which shelter it.

Some scientists have subdivided this Zone of Polygonal Wetlands into arctic and subarctic subzones that also follow a north-south gradient. The primary differences are in climate, vegetation, and thickness of peat deposits, with thicker peat to the south. Vegetation forms are similar, but the species mix is different as one moves south, and dwarf shrubs increase in frequency. One distinctive plant that occurs in the southern part of the Zone of Polygonal Wetlands is the low-growing cloudberry. The subarctic zone has fewer polygonal bogs, and more homogeneous fens and bogs than the arctic zone.

Fauna of the polygonal wetland zone must be adapted to the extreme conditions. Soil biota consists of only a few species able to cope with low-oxygen and acidic conditions, mainly springtails (small insects of order Collembola), mites, and small worms called enchytraeids. True flies (Diptera), beetles (Coleoptera), true bugs (Heteroptera), and aphids, leafhoppers, and scales (Homoptera) appear in great numbers during the warm months. Indeed, so numerous are the biting insects (mosquitoes, midges, gnats) that travelers find it nearly impossible to visit the area in the early summer. Spiders (order Araneae) are numerous and diverse as well. Reindeer, elk, and many birds, particularly ducks, geese, and wading birds, use this area seasonally. A species of great importance in the food web is the lemming, whose great numbers support the red fox, arctic fox, and wolf, as well as several raptors. These carnivores feed on the mountain hare as well. Polar bears are sometimes seen near the coast. Many of the mammals have adapted to the changing conditions by changing colors: the arctic fox, the lemming, and mountain hare develop white fur during the winter. This strategy helps to minimize radiant heat loss as well as provide camouflage.

The Palsa Zone is so-called because of the occurrence there of palsas, landforms created by the formation of ice lenses or domes. Covered with peat, these ice lenses are a form of permafrost whose continued existence is made possible by the insulating layer of peat. In the Palsa Zone of the WSL, they are 6–12 ft (2–4 m) high in the northern part of the zone and 18–24 ft (6–8 m) high in the southern part. Their diameter is typically 60–300 ft (20–100 m).

This zone lies primarily in the lowland between the Ob and Yenisei Rivers. Palsa wetlands cover about half of the zone, the northern part of which is underlain by continuous permafrost and the southern by sporadic permafrost. The palsas are covered by dwarf shrubs, cloudberry and other forbs, and mosses; low hollows between them are dominated by sedges, cotton grasses, herbs, and mosses.

This region is an ecotone between the tundra and the tiaga, so its faunal diversity is relatively high. Animals include many of those seen in the Zone of Polygonal Bogs, but otters appear, as do the highly endangered West Siberian beaver and Siberian crane. A number of migratory birds summer here as well, including several more southerly species not found in the Zone of Polygonal Bogs.

The vast Zone of String Bogs, sometimes called the West Siberian Bog Zone, covers almost 0.5 million mi^2 (nearly 1.3 million km^2) of which 50–75 percent is peat bog. It includes the floodplains of the Ob and Irtysh, and the flat lands separating the two rivers. Although the landscape shows considerable variety, much of it consists of convex, raised bogs, the centers of which may be as much as 30 ft (10 m) higher than the edges. Typically, the centers consist of oligotrophic, treeless wet bogs with numerous lakes, dominated by Baltic bog moss. The sloping edges are drier and are dominated by rusty peat moss together with tiaga forest stands that include Siberian pine and Scots pine. These raised bogs may be several miles in diameter.

A major feature of this region is the appearance of string bogs (also called ladder bogs, aapa peatlands, patterned fens, and string fens). They often form in areas where there is a strong directional flow, however slow it may be. Strings, or linear raised hummocks of peat and peat vegetation, run perpendicular to the direction of flow and appear as a series of parallel ridges or "strings," separated by pools or low wet bogs. Plant associations reflect the differences in elevation and moisture between the ridges and the low-lying areas in between. The strings may be about 10 ft (3 m) wide and 30 ft (10 m) apart.

Upland areas in this zone are covered by boreal forests or paludified string bogs and raised bogs. Floodplains and low-lying areas tend to be either forested or wet meadow/shrub fens and marshes. One feature of note is the huge Vasyugan wetlands complex, at 3,800 mi^2 (10,000 km^2), reputedly the largest wetland in the world. It consists primarily of raised bogs and string bogs along the Vasyugan River and its tributaries. The Vasyugan has the ignominious distinction (through no fault of its own) of being the site where thousands of Stalin's internal exiles were left to die.

The Zone of String Bogs supports a diversity of animal life, including thousands of insect species, some of which reach huge population levels during the warm months. Migratory birds winging their way to and from the tundra's breeding grounds consume vast quantities of insects. Mammals include the brown bear, often called grizzly bears in North America; the lynx; the extraordinarily strong and aggressive wolverine; and several of the wolverine's relatives (members of the family Mustelidae), including otters, sables, martens, ermines, weasels, and mink.

The Pine Bog Zone is essentially a strip 90–100 mi thick (roughly 150 km) from north to south and 1,200 mi (about 2,000 km) across, which covers an area larger than Colorado. About 20 percent of it is peatland; the dry areas are covered by deciduous aspen-birch forests and, in the north, by the southern tiaga forests. A mix of different bog and fen types is found, from oligotrophic raised bogs to meso-trophic and eutrophic sedge-moss fens. Wet meadows in floodplains are dominated by mosses and sedges, as well as grasses and reeds. Woody swamps with pine and European white birch and a groundcover of grasses and sedges are abundant.

The Fen Zone is a large region (170,000 mi^2 or 440,000 km^2; about the size of Washington and Oregon combined) that is higher, drier, and warmer than the zones to the north. It marks the transition from the WSL to the steppe. The land is flat and poorly drained; many of its rivers end in lakes without outlets. Some are fresh, some are saline. Climate is continental, with an average of only about 15 in (390 mm) of precipitation per year. Winters are bitter cold, summers are hot, and droughts are common. Vegetation consists of grasslands and birch-aspen forests. Peatlands, which cover only about 5 percent of the area, are mainly eutrophic sedge-moss, reed, and grass communities dominated by common reed, slimstem reedgrass, and common rivergrass along with other grasses, sedges, and herbs. Rel-atively few of the wetlands are oligotrophic and mesotrophic bogs and fens; these persist intermingled among the eutrophic reed marshes as vestiges of earlier climate regimes.

Mammals of this region include a number of small ground-dwelling rodents, including the hamster. The wolf lives on these rodents, and takes the occasional roe deer as well. The region's wetlands are extremely important habitat for migra-tory waterbirds.

The future of the WSL. There are two major threats to the integrity of the ecosys-tems of the WSL: (1) energy exploration and production, and (2) global climate change. Agriculture has affected some areas, and logging has altered forest compo-sition, particularly in the south. Hunting and poaching have taken a toll on some species. But oil and gas production, with its associated pollution and fragmentation of habitat, has had the greater impact, particularly in the north. The effects of global climate change are predicted to be greater at the high latitudes, and the pre-dicted changes seem to be happening, with a trend toward rising temperatures and melting permafrost. A great outpouring of published research focuses on the peat-lands of the WSL as a sink for global carbon, and as a possible source of green-house gases, particularly methane, as the region warms.

Human Impacts on Wetlands

The human impact on wetlands has been extensive and pervasive. According to Mitsch and Gosselink (2000, p. 38), it is "probably safe to assume that we are still losing wetlands at a fairly rapid rate globally and that we have lost as much as 50 percent of the original wetlands on the face of the Earth." Wetlands, long seen as waste places, have been filled, drained, and buried under water behind dams.

Common Human Impacts on Wetlands

Common human impacts on wetlands include hydrologic alteration, pollution, disturbance, and fragmentation.

Hydrologic Alteration

- Stabilization of water levels can reduce or eliminate disturbance (flood pulse) required to maintain the system; often a consequence of dam operations, particularly those aimed at flood control.
- Increased water levels or lengthened hydroperiod; may be caused by impoundments or land subsidence resulting from oil and gas extraction.
- Decreased water levels or shortened hydroperiod; typically caused by drainage, often by ditching for agriculture, or by increased water extraction.
- Decreased water levels caused by elevating ground surface, that is, filling in wetlands.

Pollution

- Increased siltation from land-disturbing activities in the watershed.
- Nutrient enrichment from point or nonpoint sources, or from atmospheric deposition.
- Increased levels of toxic substances, such as metals or pesticides.

Disturbance

- Setting fires, or suppressing fires in systems requiring periodic fires.
- Off-road vehicle use.
- Logging.
- Introduction of invasive exotic species, such as common reed or nutria.
- Removal of species through hunting or poaching, leading to cascading alterations of trophic relationships, species composition, and even physical habitat as in the case of beavers.

Fragmentation

- Construction of roads, canals, drainage ditches, and levees that interrupt hydrologic processes and block species movement and dispersal.

Where not intentionally converted to agricultural or urban land, they have been fragmented by roads and canals, and degraded through logging, water pollution, hydrologic alteration, and introduction of exotic invasive species. These processes of wetlands destruction are well established in industrial countries and, as the less-developed countries adopt the ways of the developed countries, they are spreading. Wetlands in the most remote places on Earth are being affected. The Pantanal is at risk from upland agriculture, tourism, and large-scale water development schemes. Wet meadows and marshes throughout the world are being converted to agricultural uses. The vast peatlands of the high-latitude north are experiencing rapid climate change, with unknown ecological repercussions. Although not discussed in this volume, mangrove swamps in some parts of the world are under pressure from the expansion of aquaculture.

What, then, does the future hold for the world's wetlands? Barring some catastrophe, it is likely that the human population will reach at least 9 billion by the end of this century, and that world economic expansion will continue. The underlying trends that historically have led to wetlands destruction will intensify. At the same time, attitudes toward wetlands have changed considerably, and wetlands conservation and restoration are being widely applied.

The value of intact, healthy wetlands is now widely recognized among scientists and resource managers (see Table 3.1). Many of the values and functions of wetlands take the character of what economists term "public goods"—economic goods whose existence benefits everyone, or at least a large number of people. In market economies, public goods tend to be neglected because the private owners of such resources cannot easily capitalize on them. The public, used to benefitting from wetlands for free, has no incentive to pay for their continued existence, and a mechanism for doing so is not obvious even if it wanted to. Therefore, a great many programs aimed at wetlands conservation and restoration are governmental.

The U.S. federal government and its programs are major drivers of wetlands conservation throughout the United States. Since the 1980s, every U.S. administration has adopted at least a goal of "no net loss" of wetlands; several, including the Clinton and George W. Bush administrations, have aimed at a net increase.

There is also an international agreement to conserve wetlands. While the original focus of the Ramsar Convention was on the conservation of wetlands for their habitat value for waterfowl, the scope of the Convention has broadened to include all aspects of wetlands conservation and use. The sustainable use of wetlands promoted by the Convention means beneficial human use that is compatible with maintenance of the natural functions and values of the wetland. The Ramsar Convention, which is administered by the United Nations, maintains a list of wetlands of international importance, known as "Ramsar sites." As of November 2006, there were 1,634 Ramsar sites on the list, covering a total area of 562,312 mi^2 (1.45 million km^2). The Ramsar Convention also conducts research and disseminates information.

Wetlands Creation and Restoration

The art and science of wetlands creation and restoration has developed largely as a result of the implementation of the goal of no net loss. Much has been learned through trial and error. Wetlands restoration is generally seen as more likely to achieve success than the creation of a wetland on a site where there was none previously. Many wetland scientists, however, are concerned that wetlands restoration and, even more, wetlands creation result in long-term loss of some functions and values of wetlands. Most restored and created wetlands have not been in existence long enough to be able to assess long-term success.

Further Readings

Books

Douglas, Marjory S. 1997. *The Everglades: River of Grass.* 50th anniversary ed. Sarasota, FL: Pineapple Press.

Lerner, Carol. 1983. *Pitcher Plants: The Elegant Insect Traps.* New York: William Morrow.

Lyons, Janet, and Sandra Jordon. 1989. *Walking the Wetlands: A Hiker's Guide to Common Plants and Animals of Marshes, Bogs, and Swamps.* 3rd ed. New York: Wiley.

Internet Sources

California Vernal Pools. n.d. "Information and Resources on Vernal Pool Wetlands in California and Beyond." http://www.vernalpools.org.

Florida Museum of Natural History. n.d. "Ichthyology at the Florida Museum of Natural History." Information on different habitats and plants and animals of the Everglades. http://www.flmnh.ufl.edu/fish/southflorida/everglades.html.

Irish Peatland Conservation Council. n.d. "Peatlands around the World." Pictures of peatlands of many countries. http://www.ipcc.ie/wptourhome1.html.

North Carolina State University Water Quality Group. n.d. "WATER SHEDSS Educational Component." Information on wetlands and aspects of water quality. http://www.water.ncsu.edu/watershedss/info/wetlands/index.html.

U.S. Department of Agriculture, Natural Resources Conservation Service. n.d. "NRCS Wetland Science Web Site." Information on hydric soils and wetlands restoration. http://www.wli.nrcs.usda.gov.

U.S. Environmental Protection Agency. n.d. "Office on Wetlands, Oceans, and Watersheds." http://www.epa.gov/owow/wetlands.

Waterland Research Institute. n.d. "Welcome to the Waterland Institute." Comprehensive information about the Pantanal and efforts to conserve it. http://www.pantanal.org.

Appendix

Selected Plants and Animals of Wetlands

Miscellaneous Wetland Biota

Plants

Speckled alder	*Alnus incana*
European alder	*Alnus glutinosa*
Red mangrove	*Rhizophora mangle*
Water lily	Family Nymphaeceae
Lotus	*Nelumbo* spp.
Common reed	*Phragmites australis*
Southern cattail	*Typhus domingensis*
European alder	*Alnus glutinosa*
Rice	*Oryza sativa*
Bald cypress	*Taxodium distinchum*

Vertebrates

Jacare	*Caiman crocodilus yacare*
American alligator	*Alligator mississippiensis*
Jabiru stork	*Jabiru mycteria*
Caribbean flamingo	*Phoenicopterus ruber*
South American jaguar	*Panthera onca*
Asian fishing cat	*Prionailurus viverrinus*

Tidal Freshwater Marshes

Plants (Low Marsh)

Spatterdock	*Nuphar advena, N. luteum*
Arrow arum	*Peltandra virginica*
Pickerelweed	*Pontederia cordata*

Plants (High Marsh)

Wild rice	*Zizania aquatica* var. *aquatica*
Calamus	*Acorus calamus*
Halberdleaf tearthumb	*Polygonum arifolium*
Jewelweed	*Impatiens capiensis*
Wild rice	*Zizania aquatica*
Cattail	*Typha latifolia*

Plants (Mid-Marsh)

Marsh smartweed	*Polygonum punctatum*
Tidalmarsh amaranth	*Amaranthus cannibina*

Plants (Floating Mats, Gulf Coast of North America)

Maidencane	*Panicum hemitomum*
Olney threesquare	*Scirpus americanus*
Cattail	*Typha latifolia*
Giant bulrush	*Scirpus californicus*

Plants (New Deltas, Gulf Coast of North America)

Willow	*Salix nigra*
Sedge	*Scirpus deltarum*
Arrowhead	*Sagittaria latifolia*
Cattail	*Typha latifolia*

Fishes (North America)

American eel	*Anguilla rostrata*
White shad	*Alosa sapidissima*
Blueback herring	*Alosa aestivalis*
White perch	*Morone americana*

Reptiles (North America)

Snapping turtle	*Chelydra serpentina*
Water moccasin	*Agkistrodon piscivorus*
American alligator	*Alligator mississippiensis*

Mammals (North America)

Marsh rice rat	*Oryzomys palustris*
Virginia opossum	*Didelphis virginiana*
Marsh rabbit	*Sylvialagus palustris*
Mink	*Mustela vison*

(Continued)

Muskrat	*Ondatra zibethicus*
River otter	*Lontra canadensis*
Raccoon	*Procyon lotor*
Nutria (introduced)	*Myocastor coypus*
White-tailed deer	*Odocoileus virginianus*

Nontidal Freshwater Marshes (Delta Marsh, Canada)

Plants

Fennel pondweed	*Potamogeton pectinatus*
Common water milfoil	*Myriophyllum sibiricum*
Common hornwort	*Ceratophyllum demersum*
Bulrush	*Scirpus acutus*
Cattail	*Typha glauca*
Common reed	*Phragmites australis*
Rivergrass	*Scolochloa festucacea*
Wheat sedge	*Carex atherodes*
Sandbar willow	*Salix interior Rowlee*

Fishes

Fathead minnow	*Pimephales promelas*
Five-finned stickleback	*Culaea inconstans*
Nine-spined stickleback	*Pungittius pungittius*
Yellow perch	*Perca flavescens*
Spot-tailed shiner	*Notropis hudsonias*
White sucker	*Catostomus commersoni*
Carp	*Cyprinus carpio*
Brown bullhead	*Ictalurus melas*
Black bullhead	*Ictalurus nebulasus*
Iowa darter	*Etheostoma exile*

Birds

Canada Goose	*Branta canadensis*
American Wigeon	*Anas americana*
Cinnamon Teal	*Anas cyanoptera*
Canvasback	*Aythya valisineria*
Green-winged Teal	*Anas crecca*
Lesser Scaup	*Aythya affinis*
Gadwall	*Anas strepera*
Blue-winged Teal	*Anas discors*
Bufflehead	*Bucephala albeola*
American Black Duck	*Anas rubripes*
Mallard	*Anas platyrhynchos*
Northern Shoveler	*Anas clypeata*

Common Goldeneye	*Bucephala clangula*
Wood Duck	*Aix sponsa*
Ring-necked Duck	*Aytha collaris*
Snow Goose	*Chen caerulescens*

Swamps

Plants (Congo River)

Bubinga	*Guibourtia demeusei*
Kratom	*Mitragyna speciosa*
Trees (no common names)	*Symphonia globulifera, Entandrophragma palustre, Symphonia globulifera, Endandrophragma palustre, Uapaca heudelotii, Sterculia subviolacea, Manilkara* spp., *Garcinia* spp.
Alstonia (buttress roots)	*Alstonia congensis*
Raphia palm	*Raphia farinifera*

Mammals (Irrawaddy Delta, Myanmar)

Asian elephant	*Elaphas maximus*
Tiger	*Panthera tigris*
Leopard	*Panthera pardus*

Peatlands

Bog Plants

Scots pine	*Pinus sylvestris*
Pitcher plant	*Sarracenia* spp.
Sundew	*Drosera* spp.
Venus flytrap	*Dionaea* spp.

Tropical Peatland Plants

Ramin	*Gonystylus bancanus*
Jongkong	*Dactylocladus stenostachys*
Alan	*Shorea albida*
Sepetir	*Copaifera palustris*

Prairie Potholes

Plants (High Zone)

Goldenrod	*Solidago* spp.
Kentucky bluegrass	*Poa pratensis*

(*Continued*)

| Giant sandreed | *Calamovilfa gigantea* |
| Smooth brome | *Bromus inermis* |

Plants (Mid-Zone)

Timothy	*Phleum pratense*
Common spikerush	*Eleocharis palustris*
Reedgrass	*Calamagrostis* spp.
Upright sedge	*Carex stricta*

Plants (Low Zone)

Reedgrass	*Calamagrostis* spp.
Wheat sedge	*Carex atherodes*
Blister sedge	*Carex vesicaria*
Water knotweed	*Polygonum amphibium*
Broadfruit burreed	*Sparganium eurycarpum*
Cattail	*Typha latifolia*

Plants (Saline)

Hardstem bulrush	*Schoenoplectus acutus*
Alkali bulrush	*Scirpus maritimus*
Greasewood	*Sarcobatus vermiculatus*
Saltbush	*Atriplex* spp.
Pickleweed	*Salicornia* spp.

Birds (Nesting)

Loons
| Common Loon | *Gavia immer* |

Grebes
Red-necked Grebe	*Podiceps grisegena*
Horned Grebe	*Podiceps auritus*
Pied-billed Grebe	*Podilymbus podiceps*

Pelicans
| American White Pelican | *Pelecanus erythrorhynchos* |

Cormorants
| Double-crested Cormorant | *Phalacrocorax auritus* |

Herons
Great Blue Heron	*Ardea herodias*
Green-backed Heron	*Butorides virescens*
Great Egret	*Ardea alba*

Black-crowned Night-heron	*Nycticorax nycticorax*
Least Bittern	*Ixobrychus exilis*
American Bittern	*Botaurus lentiginosus*

Waterfowl

Tundra Swan	*Cygnus columbianus*
Canada Goose	*Branta canadensis*
Snow Goose	*Chen caerulescens*
Mallard	*Anas platyrhynchos*
Gadwall	*Mareca strepera*
Northern Pintail	*Anas acuta*
Green-winged Teal	*Anas carolinensis*
Blue-winged Teal	*Anas discors*
American Wigeon	*Mareca americana*
Northern Shoveler	*Anas clypeata*
Wood Duck	*Aix sponsa*
Redhead	*Aythya americana*
Ring-necked Duck	*Aythya collaris*
Canvasback	*Aythya valisineria*
Lesser Scaup	*Aythya affinis*
Common Goldeneye	*Bucephala clangula*
Bufflehead	*Bucephala albeola*
Ruddy Duck	*Oxyura jamaicensis*
Common Merganser	*Mergus merganser*

Cranes

Sandhill Crane	*Grus canadensis*

Rails

Virginia Rail	*Rallus limicola*
Sora	*Porzana carolina*
American Coot	*Fulica americana*

Plovers

American Avocet	*Recurvirostra americana*
Semipalmated Plover	*Charadrius semipalmatus*
Killdeer	*Charadrius vociferus*

Sandpipers

Ruddy Turnstone	*Arenaria interpres*
American Woodcock	*Scolopax minor*
Common Snipe	*Gallinago gallinago*
Spotted Sandpiper	*Actitis macularia*
Solitary Sandpiper	*Tringa solitaria*
Greater Yellowlegs	*Tringa melanoleuca*
Lesser Yellowlegs	*Tringa flavipes*

(*Continued*)

Pectoral Sandpiper	*Calidris melanotos*
White-rumped Sandpiper	*Calidris fuscicollis*
Baird's Sandpiper	*Calidris bairdii*
Least Sandpiper	*Calidris minutilla*
Dunlin	*Calidris alpina*
Short-billed Dowitcher	*Limnodromus griseus*
Stilt Sandpiper	*Calidris himantopus*
Semipalmated Sandpiper	*Calidris pusilla*
Marbled Godwit	*Limosa fedoa*
Wilson's Phalarope	*Phalaropus tricolor*

Gulls and Terns

Herring Gull	*Larus argentatus*
Ring-billed Gull	*Larus delawarensis*
Franklin's Gull	*Larus pipixcan*
Forster's Tern	*Sterna forsteri*
Black Tern	*Chlidonias niger*

Playa Lakes

Plants (Short Hydroperiod)

Wheatgrass	*Pascopyrum smithii*
Buffalograss	*Buchloe dactyloides*
Vine mesquite	*Panicum obtusum*
Barnyard grass	*Echinochloa crus-galli*
Pale spikerush	*Eleocharis macrostachya*
Mucronate spangletop	*Leptochloa filiformis*
Smartweed	*Polygonum* spp.

Shorebirds

American Avocet	*Recurvirostra americana*
Long-billed Dowitcher	*Limnodromus scolopaceus*
Wilson's Phalarope	*Phalaropus tricolor*
Long-billed Curlew	*Numenius americanus*
Stilt Sandpiper	*Calidris himantopus*
Lesser Yellowlegs	*Tringa flavipes*
Snowy Plover	*Charadrius alexandrinus*

The Pantanal

Plants (Floating Marshes)

| Water hyacinth | *Eichhornia* spp. |
| Eared watermoss | *Salvinia auriculata* |

Cuban bulrush	*Pistia stratiotes*
Horsetail paspalum	*Paspalum repens*
Water lettuce	*Pistia stratiotes*
Giant flatsedge	*Cyperus giganteus*
Softstem bulrush	*Scirpus validus*
Southern cattail	*Typha domingensis*

Plants (Seasonally Flooded Grasslands)

Carpetgrass	*Axonopus purpusii*
Spikerush	*Eleocharis acutangula*
Panicgrass	*Panicum* spp.
Caronal grass	*Elyonurus muticus*
Trumpet-tree	*Tabebuia aurea*
Palms (palm savannah)	*Copernicia alba, C. australis*
Annona	*Annona* spp.
Blackrodwood	*Eugenia biflora*

Plants (High Ground)

Cohoba	*Piptadenia peregrina* spp.
Cottontree	*Bombax* spp.
Guajava	*Psidium persicifolium*
Acacia	*Acacia* spp.
Kapok tree	*Ceiba pentendra*
Earpod tree	*Enterolobium contortisiliquum*
Mimosa	*Mimosa* spp.
Piptadenia	*Piptadenia* spp.
Guayacan	*Caesalpinea paraguariensis*

Fishes

Pacu caranha	*Piaractus mesoptamicus*
Curimbata	*Prochilodus lineatus*
Tambaqui	*Colossoma macropomum*
Freshwater flounders	*Hypoclinemus* spp.

Reptiles

Yellow anaconda	*Eunectes notaeus*
Boa constrictor	*Boa constrictor*
Pantanal coral snake	*Micrurus tricolor*
Neotropical rattlesnake	*Crotalus durissus*
Brazilian lancehead	*Bothrops moojeni*
Neuwid lancehead	*Bothrops neuwiedi*
Caiman	*Crocodilus yacare*

Birds

Jabiru (Tuiuiu)	*Jabiru micteria*
Hummingbirds	Family Trochilidae
Rhea	*Rhea americana*
Crested Caracara	*Polyborus plancus*
Hyacinth Macaw	*Anodorhynchus hyacinthinus*
Yellow-billed Cardinal	*Paroaria capitata*
Mato Grosso Antbird	*Cercomacra melanaria*

Mammals

Capybara	*Hydrochoeris hydrochaeris*
Swamp deer	*Blastocerus dichotomus*
Collared peccary	*Tayassu tajacu*
White-lipped peccary	*Tayassu albirostris*
Maned wolf	*Chrysocyon brachyurus*
Jaguar	*Panthera onca*
Puma	*Puma concolor*
Jaguarundi	*Herpailurus yaguarondi eyra*
Tiger cat	*Leopardus tigrinus*
Ocelot	*Leopardus pardalis*
Giant otter	*Pteronura brasiliensis*
Howler monkey	*Alouatta caraya*
Brown capuchin	*Cebus apella*

The West Siberian Lowlands

Plants (Zone of Polygonal Wetlands)

Water sedge	*Carex stans*
Polar willow	*Salix polaris*
Heath	*Ericaceae* spp.
Cloudberry	*Rubus chamaemorus* L.

Mammals (Zone of Polygonal Wetlands)

Reindeer	*Rangifer tarandus*
Elk	*Alces alces*
Lemming	*Lemmus sibiricus, Dicrostonyx torquatus*
Red fox	*Vulpes vulpes*
Arctic fox	*Alopex lagopus*
Wolf	*Canis lupus*
Mountain hare	*Lepus timidus*
Polar bear	*Ursus maritimus*

Birds (Palsa Zone)

Siberian crane	*Grus leucogeranus*

Mammals (Palsa Zone)

Otter	*Lutra lutra*
West Siberian beaver	*Castor fiber pohlei*

Plants (Zone of String Bogs)

Baltic bog moss	*Sphagnum balticum*
Rusty peat moss	*Sphagnum fuscum*
Siberian pine	*Pinus sibirica*
Scots pine	*Pinus sylvestrus*

Mammals (Zone of String Bogs)

Brown bear (grizzly bear)	*Ursus arctos*
Lynx	*Felis lynx*
Wolverine	*Gulo gulo*

Plants (Pine Bog Zone)

Reed	*Phragmites australis*
European white birch	*Betula pendula*

Plants (Fen Zone)

Reed	*Phragmites communis*
Slimstem reedgrass	*Calamagrostis neglecta*
Common rivergrass	*Scolochloa festucacea*

Mammals (Fen Zone)

Hamster	*Cricetus cricetus*
Wolf	*Canis lupus*
Roe deer	*Capreolus capreolus*

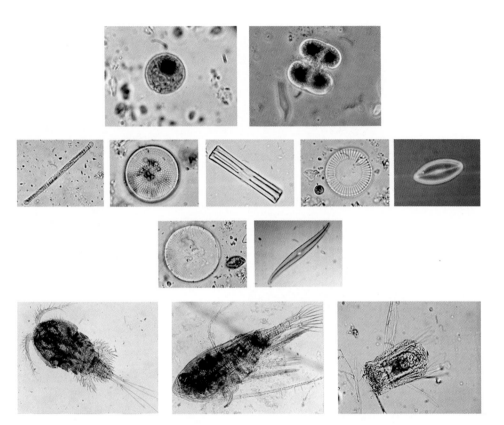

Plate I. Typical freshwater plankton: (top row, left to right) The green alga Chlamydomonas and the desmid alga Cosmarium. Diatoms: (second row, left to right) Diatom elongatum, *Coscinodiscus* spp., *Tabellaria fenestrata*, *Cyclotella meneghiniana*, *Navicula miniscula*. (third row, left to right) *Stephanodiscus niagarae*, *Gyrosigma strigile*. (fourth row, left to right) The copepod crustaceans Cyclops and Diaptomus, and the rotifer Polyarthra. *(Images from U.S. Environmental Protection Agency, Great Lakes Program Web site.)*

Plate II. Typical birds associated with freshwater habitats in North America: (clockwise from top left) American Dipper, Hooded Mergansers, Belted Kingfisher, Red-necked Grebes, Double-crested Cormorant, and Great Blue Heron. *(Images from U.S. Fish and Wildlife Service Digital Media Library except for Great Blue Heron, courtesy of Susan L. Woodward.)*

Plate III. (top) The New River's endemic candy darter. *(Photo by Noel Burkhead, USGS, Florida Integrated Science Center.)* (bottom) The famous snail darter of the Tennessee River system. *(Courtesy of U.S. Fish and Wildlife Service.)*

Plate IV. Rare and endangered freshwater mussels from Virginia's Clinch River, part of the upper Tennessee River system: (clockwise from top left) *Epioblasma brevidens*, *E. capsaeformis* with rotating microlures presumably mimicking the cercae (tails) of aquatic insects, *E. capsaeformis* capturing and inoculating a host fish, and *Lemiox rimosus* mimicking a small aquatic snail. Photos illustrate a variety of reproductive strategies in which the mussels display "mantle-lures" (an extension of their soft body or mantle) designed to attract host fish species (see Figure 2.13). *(Courtesy of Jess Jones, U.S. Fish and Wildlife Service, and Nick King, Virginia Tech.)*

Plate V. Freshwater wetlands: (clockwise from top left) High marsh in a Virginia freshwater tidal marsh, dominated by wild rice and pickerelweed. *(Photo © Irvine Wilson.)* Low marsh in a Virginia freshwater tidal marsh, dominated by spatterdock. *(Photo © Gary Fleming, Virginia Department of Conservation and Recreation.)* Bog aerial photo; cypress swamp; Great Dismal Swamp, Virginia; fringe marsh, Klamath Lake. *(Images from U.S. Fish and Wildlife Service Digital Media Library.)*

Plate VI. Plants of Cranberry Glade, a high-altitude bog in West Virginia, showing adaptations to both saturated conditions and low nutrient levels. (top, left to right) Labrador tea, small cranberry, cottongrass, and sphagnum moss with rose pogonia inset. (bottom, left to right) Pitcher plant leaves and flower; round-leaved sundew. *(Courtesy of Susan L. Woodward.)*

Plate VII. The Pantanal: (clockwise from top left) Jabiru Stork; typical Pantanal land-scape; Red-legged Sereima; Giant otters; Jacaré or spectacled caiman; Lesser Anteater; Hyacinth Macaws. (center) Wattled Jacana. *(Jabiru Stork, Lesser Anteater, Hyacinth Macaws, and Wattled Jacana by permission of Roy Slovenko; Red-legged Sereima, Giant otters, Jacaré, and typical Pantanal landscape by permission of Christopher Reiger.)*

Plate VIII. A few of the myriad of Lake Victoria cichlids: (left, top to bottom) Hippo Point blue bar (*Haplochromis* sp.); *Paralabidochromis chromogynos* (no common name); Dwarf Victoria mouthbrooder (*Pseudocrenilabrus multicolor victoriae*); Flameback (*Haplochromis* sp.). (right, top to bottom) Finebar scraper (*Haplochromis* sp.); Blue rock kribensis (*Haplochromis* sp.); Blue rock kribensis (color variant); Yellow rock kribensis (*Haplochromis* sp.). *(Courtesy of Kevin Bauman, http://african-cichlid.com.)*

4

Lakes and Reservoirs

Introduction

Lakes, ponds, and reservoirs are bodies of water characterized by relatively deep, open water that is without, or nearly without, current. There is no universally accepted, scientifically based distinction between lakes and ponds, but ponds are small and lakes are large. Since they are ecologically similar to small lakes, ponds will be subsumed under the discussion of lakes.

Reservoirs, or impoundments, are manmade lakes. In many regions of the world, they outnumber natural lakes. Reservoirs are, with some exceptions, lentic (nonflowing) bodies of water created by the building of a dam on a river. Ecologically they are quite similar to natural lakes of like proportions, except that their hydrology is subject to considerable human manipulation. They will be discussed separately from lakes.

Ways to Classify Lakes

There are a number of ways to classify lakes. They can be classified according to their origins, their nutrient status, their mixing regime, their ecoregional setting, their size and shape (whether they are on balance autotrophic or heterotrophic), and their degree of human impact.

Classification by Origins of Lakes

Natural lakes make their appearance on the landscape through a number of quite distinct processes. Many lakes are glacial in origin. The Earth has gone through a great many periods of glacial advance and retreat; these periods of heightened glacial activity are known as glaciations or Ice Ages. The most recent Ice Age ended less than 20,000 years ago, and vestiges of it still remain in the polar regions and at high altitudes. During these periods, enormous glaciers covered large parts of the Northern Hemisphere. The ice formed (and continues to form) lakes in several ways. A large glacier flowing down a valley or across a broader landscape pushes before it a large amount of material, which may remain as a natural dam to stream flow when the glacier retreats. A huge sheet of ice moving across the landscape leaves behind it scoured-out depressions that fill with meltwater and rainwater. The Great Lakes of North America formed in that way, as did many smaller lakes in the Northern Hemisphere. Retreating glaciers may leave behind massive blocks of ice embedded in the ground, and when these melt, lakes form in the resulting depressions.

Some of the largest, oldest, and deepest lakes in the world were formed by tectonic movement of the Earth's surface. Troughs called rift valleys formed when a section of crust material either subsided (sank) relative to its surroundings, or the surroundings rose or simply pulled apart along a fault line. Rift valleys are some of the most spectacular landscapes on Earth. The Great Rift Valley of Africa extends from the central eastern part of that continent some 3,000 mi (4,800 km) north into Syria. This valley contains two of the world's largest lakes, Lake Tanganyika and Lake Malawi, each hundreds of miles long and very deep. Lake Baikal, an Asian rift valley lake, is the world's oldest, deepest, and largest (by volume) lake.

Lakes are also formed by another phenomenon associated with tectonic activity—volcanoes. The cones on volcanoes can collapse after eruption, or an even larger part of the volcano can collapse back into itself, forming a depression known as a *caldera*. If the depression fills with water, a lake forms. Crater Lake in Oregon is a good example of this type of lake.

Many lakes are associated with rivers. As large rivers meander across their floodplains, they form and reform channels, cutting through old meander curves and abandoning former channels. These old channels, now curved lakes, are known as oxbow lakes or, in Australia, as billabongs. Furthermore, the eroding, sorting, and redepositing of material on the floodplain can result in the formation of other floodplain lakes. Floodplain lakes, whether oxbow or not, are periodically reunited hydrologically with the river during high water events. Lakes also form along coastlines, in situations in which a coastal dune or barrier island has formed, and streams fill the depression behind it with fresh water.

Lakes that have existed continuously for very long periods are known as "ancient lakes." One source uses 100,000 years of continuous existence as a minimum for qualifying as an ancient lake; others set the bar higher. The biota of such lakes has had plenty of time for speciation to occur, and such lakes may harbor many

Table 4.1 Ancient Lakes of the World

LAKE	SURROUNDING COUNTRIES	ESTIMATED AGE (MILLIONS OF YEARS)
Eyre	Australia	20–50
Maracaibo	Venezuela	>36
Issyk-Kul	Kyrgyzstan	25
Baikal	Russian Federation	20
Tanganyika	Tanzania, Burundi, Zaire, Zambia	9–20
Caspian Sea	Iran, Kazakhstan	>5
Aral Sea	Kazakhstan, Uzbekistan	>5
Ohrid	Albania, Macedonia	>5
Prespa	Albania, Greece, Macedonia	>5
Lanao	The Philippines	3.6–5.5
Titicaca	Bolivia, Peru	3
Malawi	Malawi, Mozambique, Tanzania	>2
Tahoe	United States	2
Khubsugul	Mongolia	1.6
Buwumtwi	Ghana	>1
Vostok	Antarctica	>1
Crater	Canada	>1
Pingualuk	Canada	>1
Victoria	Kenya, Tanzania, Uganda	12,000–750,000?
Biwa	Japan	>0.4

Sources: Adapted from Duker and Borre 2001; with data from Lerman, Imobden, and Gat 1995; Groombridge and Jenkins 1998; and Rossiter and Kawanabe 2000.

endemic species. The number of ancient lakes is relatively small, 20 by some counts (the criteria are not precise, nor are the age estimates, so considerable disagreement occurs in the literature). Most are rift valley or tectonic lakes, although Lake Busumtwi's depression (in Ghana) was created by a meteor impact. The 20 oldest lakes are shown in Table 4.1.

Classification by Nutrient Status

The classification of lakes according to their nutrient status focuses on the biotic community in a lake, particularly the plants. Aquatic and semiaquatic plants, particularly the algae, are highly responsive to nutrient levels. Nutrients are chemicals needed by plants and animals alike: the chemical building blocks of life. Plants require nitrogen, potassium, calcium, magnesium, phosphorus, and sulfur, in descending order of quantity required, to build carbohydrates and more complex molecules. They also require a host of other chemicals in small quantities; these are termed micronutrients.

In lakes (and bodies of water in general), plants usually have plenty of all the nutrients they need except for nitrogen and phosphorus. A fundamental law of biology is that populations (of plants, animals, bacteria, or any living organism) grow until they run out of some key nutrient: this is Liebig's Law of the Minimum. Liebig's law is often generalized to say that populations will grow until they run out of any necessary factor, for example, light. The lack of some necessary nutrient or physical factor is not the only thing that limits population growth, however. Herbivory (consumption of plants by herbivores), predation (consumption of animals by other animals), and disease often limit populations. The key nutrients that limit population growth are termed limiting nutrients; nitrogen, phosphorus, or both are frequently limiting nutrients in aquatic systems.

Although there are no generally agreed-upon, precisely measurable differences between lakes with different nutrient levels (see Table 4.2), a lake is termed oligotrophic (from Greek meaning "low level nutritious") if its low level of key nutrients

Table 4.2 Characteristics of Lakes of Differing Nutrient Status

	Oligotrophic	Mesotrophic	Eutrophic	Dystrophic
Water clarity	High	Medium	Low due to high plankton concentration	Often low due to humic acid coloration but sometimes high
Phosphorus level	Relatively low	Medium	Relatively high	Low
Nitrogen level	Relatively low	Medium	Relatively high	Low
Bottom material	Rocky, little sediment accumulation	Some sediment accumulation	Considerable accumulation of sediment and organic material	Consists almost entirely of organic material (peat)
Planktonic population level	Low, little production	Moderate	High to very high	Generally low, with low species diversity
Macrophyte population level	Low, few aquatic plants despite abundant light in littoral zone	Moderate	Littoral zone supports extensive aquatic and emergent plant populations	High but few species, may be dominated by sphagnum and other mosses
Dissolved oxygen level	High, often throughout the water column	Moderate; may be stratified	Low, particularly at depth; diel fluctuations	Low
Fishery type	Cold water	Warm water	Warm water	Cold water, but low pH may preclude fish

results in a generally low level of plant biomass, particularly of algae. A eutrophic (meaning "high level nutritious") lake, at the other end of the spectrum, typically is rich with plant and animal life. Often in extreme cases, for example, where human wastewater discharges to a lake have greatly increased nitrogen and phosphorus levels, lakes are termed "hypereutrophic." Mesotrophic ("medium level nutritious") lakes occupy the middle ground between oligotrophic and eutrophic lakes. A last category is sometime included called a dystrophic ("malnutritious") lake. While oligo-, meso-, and eutrophic all refer to the level of plant production within the lake itself, dystrophic lakes have low levels of production but high levels of organic (carbon-based) material. The organic material is of terrestrial origin (leaves, pine needles) and may be in dissolved form. Lakes classified as dystrophic are usually highly acidic bog lakes dominated by sphagnum moss (see Chapter 3).

Autotrophic and Heterotrophic Lakes

Lakes are called autotrophic if their food webs are based primarily on plant production in the lake, and heterotrophic if the greater part of energy available to their food webs is from organic material supplied to the lake from terrestrial sources. The terrestrial organic material may take the form of leaves and pine needles; excreta from terrestrial animals; seeds, fruits, and pollen; and dissolved organic compounds from organic material in the soils. In autotrophic lakes, the amount of carbon stored through photosynthesis is greater than that used by the lake biotic community in respiration. In heterotrophic lakes, the greater part of food energy that moves through the lake food web comes from the decomposers, which make organic material of terrestrial origin available to the lake food web. In the net-heterotrophic lake, more energy is used in respiration than is produced in photosynthesis in the lake. Taken together, the world's lakes are heterotrophic, putting more carbon dioxide into the atmosphere via respiration than they remove via photosynthesis.

Classification Based on Mixing

Lakes are sometimes classified according to how often their waters are mixed.

A characteristic of many lakes, particularly deeper lakes, is thermal stratification, the separation of water into layers that infrequently, or never, mix. The lowest level in a stratified lake is called the hypolimnion, and it is frequently characterized by very low oxygen levels, which make it difficult for many organisms to exist therein. Thus, the frequency and throughness of mixing are important for lake ecology, for it is through mixing that deeper waters are made habitable for fish and other biota.

Mixing often takes place as a result of wind, and therefore is often seasonal. In the temperate regions, windy conditions are usually associated with spring and fall, while summer is a time when wind is lacking. Mixing is also a function of seasonality in that the warming (in spring and summer) and the cooling (in fall and winter) of the top layer of water make it the same density as the bottom layer on more than one occasion during the course of the year, and mixing can then occur. Finally, the size and shape of the lake also contribute to its mixing regime: large, shallow lakes are more subject to the influence of the wind than small, deep lakes, particularly those sheltered by trees, hills, or bluffs. In fact all lakes, particularly large lakes, are much more subject to wind influences than typical lake diagrams in books might lead one to believe, for the vertical (depth) scale is usually greatly exaggerated. Consider Lake Baikal, the world's deepest lake. It is 395 mi (636 km) long, 50 mi (80 km) wide, and just over 1 mi (1,637 m) deep. In other words, it is almost 400 times as long as it is deep, and its proportions are like those of a ditch 1 in (2.54 cm) deep at its maximum, 4 ft (1.22 m) wide, and 33 ft (10 m) long. Clearly, such a lake will have a great deal of surface area in relation to its volume of water, giving the winds plenty of influence.

Different lakes have different mixing regimes, since the variables that influence mixing—climate, size, and shape—are different for different lakes. The categories of lakes according to their mixing regime are as follows:

- Monomictic: lakes that mix from top to bottom once a year, usually for a relatively brief period
- Dimictic: lakes that mix from top to bottom twice a year (common in temperate-zone lakes), once in the spring, once in the fall
- Polymictic: lakes that stratify and then mix a number of times each year
- Oligomictic: lakes that rarely if ever mix; such lakes are common in the tropics
- Meromictic: lakes that may mix in the upper layers but have a bottom layer that rarely if ever mixes, partly because of its low temperature and therefore high density and partly because of its higher concentrations of dissolved solids (which also increase density).

An interesting (and occasionally deadly) phenomenon associated with a few meromictic lakes is the limnic eruption: the rapid mixing of deep waters hypersaturated with dissolved gases, especially carbon dioxide. When this occurs, sometimes as a result of a landslide or other physical disturbance, a large-scale release of the gas may occur: the lake, in effect, burps. However, this can be a serious matter, as was the case with Lake Nyos, a volcanic lake in the west African country of Cameroon. In 1986, for reasons not fully understood, a limnic eruption took place releasing a large cloud of carbon dioxide mixed with other gases. The gas suffocated almost 2,000 rural inhabitants and their livestock. A similar carbon dioxide release took place at Lake Monoun, also in Cameroon, in 1984.

Classification by Degree of Human Impact

Some lakes far from centers of industrial activity are referred to as "pristine" lakes. Such a designation is not so much a classification of lakes as a recognition that there is a spectrum of human influence, from lakes relatively lightly touched to those fundamentally altered. There probably is no such thing as a pristine lake, if only because of the pervasive and ubiquitous human impacts on climate and atmospheric chemistry. No region on Earth is untouched. At the other end of the spectrum are lakes such as the Aral Sea, described later, which have been catastrophically affected by human influence. Human-created lakes are also at this end of the spectrum.

The Physical Template

The physical environment of lakes is the stage upon which the living organisms of lakes live their lives and to whose characteristics they must adapt or die. To the untrained eye, it might seem that lakes are relatively simple, homogeneous environments. But in fact the lake environment is complex and varied, with many physical variables creating a diversity of environments.

Hydrology

At the most basic level, a lake is a depression in the land surface filled with water. Its volume is the amount of water it can hold, and this is the product of the lake's length times its width times its volume. Water is periodically or continually added to the lake, and water more or less continually leaves the lake. The inputs are small in number: precipitation falling directly on the lake's surface; groundwater flowing into the lake; and surface water, in streams and rivers, flowing into the lake. Outputs are also small in number: surface water flowing out of the lake through an outlet stream or river; evaporation (quite significant in hot dry regions); and leakage into faults or general discharge into the groundwater system.

A constant lake level (that is, water surface elevation) is the result of the sum of the water inputs equaling the sum of the water outputs. But lake levels are hardly ever constant for long. Precipitation may be seasonal, causing the lake level to rise during the wet season and fall during the dry season. If the lake has an outlet (those that do not are called endorheic lakes), the rate of flow will increase when the lake level rises and decrease as it falls. Evaporation is often seasonal, depending on solar intensity, wind, and relative humidity as well as lake water temperature.

The rate of flow into and out of a lake in relation to its volume determines the retention time or residence time. This is the average amount of time that any particular water molecule will spend in the lake. It is determined by dividing the volume of water in the lake by the rate of outflow. Residence times vary considerably from lake to lake. Lake Tahoe on the California-Nevada border has a residence time of about 700 years. This has clear implications for the accumulation of nutrients and pollutants and thus for water quality. The longer the residence time, the longer such substances will stay in the lake rather than be "flushed out."

Water Chemistry

For a general discussion of such water chemistry topics as pH, dissolved solids, and dissolved oxygen, see Chapter 1.

Temperature, Density, Stratification, and Mixing

In deeper tropical lakes, and in temperate lakes in summer, it is common for the water to separate into a warmer layer near the surface, a cooler layer near the bottom, and between them a layer of intermediate temperature and density. This phenomenon, called stratification, is of great importance for lake ecology and is usually caused by density differences between warm water and cooler water. Water density, defined as mass per unit volume of water, reaches its maximum at 39.2° F (4° C); water either colder or warmer than that is less dense. At 32° F (0° C), water freezes. Ice floats because it is less dense than cold water. As the temperature increases above 39.2° F, density decreases again, causing the warmer water to "float" on top of the cooler water.

In stratified lakes, temperature does not usually decrease with increasing depth at a constant rate. There is (usually in summer in temperate regions) a particular depth at which there is an abrupt shift to cooler water. This relatively thin stratum (layer) is known as the *thermocline*. The thermocline lies in the intermediate temperature-density zone (the metalimnion); above it lies the warmer epilimnion, and below it is the cooler hypolimnion (see Figure 4.1).

In the hypolimnion, rates of oxygen-producing photosynthesis are low or nil, but a constant "snow" of organic debris settles down from the lake layers above: dead fish and other organisms, dead phyto- and zooplankton, leaf litter, pollen, and feces from organisms of all sizes. Bacteria decompose this material, using oxygen in the process. When oxygen is used up, it is not replenished. Decomposition continues, performed by bacteria that can actively metabolize organic material in the absence of oxygen ("anaerobic" bacteria).

Stratification affects not only dissolved oxygen levels but nutrient levels as well. In the epilimnion, algae take up the available nutrients. These nutrients are removed from the epilimnion as algae die and sink, or are eaten by zooplankton whose feces sink. Thus, the hypolimnion, though depleted of oxygen, is rich in nutrients; while in the epilimnion, biological production may be limited by lack of nutrients. The mixing of lake waters is thus critical not only to supply oxygen to

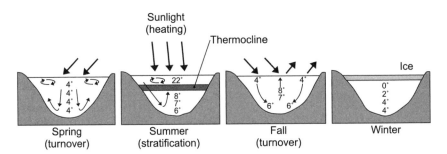

Figure 4.1 Thermal stratification and turnover (mixing) in a dimictic lake. *(Illustration by Jeff Dixon. Adapted from U.S. EPA Great Lakes Program 1995.)*

the depths but to restore nutrients to the uppermost zone, in which there is sufficient light for biological production to occur.

In the temperate zone, lakes are typically dimictic, meaning that lake waters
mix through the full depth of the water column twice a year. In winter, hypolimnial water is about 39.2° F (4° C), and the water above is cooler (though not
much cooler if it remains liquid) and therefore less dense, so there is weak stratification. In spring, as ice melts and the epilimnial water warms, there is a point
at which the temperature difference between layers becomes very small. At this
point, mixing can easily occur with a little wind. The lake "turns over." In
summer, stratification develops, with warm water in the epilimnion staying
completely separate from the cooler water in the hypolimnion. Anoxic (zero oxygen) conditions can develop in the hypolimnion. In fall, the epilimnion cools
and becomes denser, and as winter approaches, the temperature in the epilimnion will approach that of the hypolimnion. The water column can then mix
freely top to bottom (or "turn over") again.

This pattern is common in temperate zone lakes with four distinct seasons. In
tropical lakes, mixing may result from evaporative cooling during the windy season. The failure of the windy season, attributed to climate change, has reduced the
mixing of Lake Tanganyika in recent years and persistent stratification is blamed
for a dramatic decrease in productivity of the lake. Deeper tropical lakes may not
mix at all. The difference in density per increment of temperature change is not
constant; at warmer temperatures (as in the tropics) lakes can stratify with only a
3.6° F (2° C) difference between the epilimnion and the hypolimnion. There are
many possible patterns of mixing that depend on local variables of climate, topography, and the size, shape, and depth of the lake.

Light

In lakes, the zone in which there is enough light for photosynthesis to occur is
called the *euphotic zone,* defined as the water column from the surface down to the
depth at which only 1 percent of the light striking the water surface remains. This is
the light level at which photosynthesis is approximately equal to respiration. The
euphotic zone is the only zone in which plants, including phytoplankton, can live.
Crater Lake in Oregon is an ultra-oligotrophic lake; its euphotic zone extends to
about 400 ft (122 m). In eutrophic lakes, with large populations of phytoplankton,
light dies out quickly with depth; the euphotic zone may be less than 2 ft (0.61 m).
See also the discussion of light in freshwaters in Chapter 1.

Tides and Seiche Movements

Ocean tides are a familiar phenomenon. Driven by the gravitation pull of the moon
and, to a lesser extent, the sun, such astronomical tides can be quite dramatic. In
lakes, astronomical tides exert an effect but it is small, measured in fractions of an
inch. Of more importance in lakes are tide-like movements of water driven by
winds and changes in atmospheric pressure. An air pressure difference associated

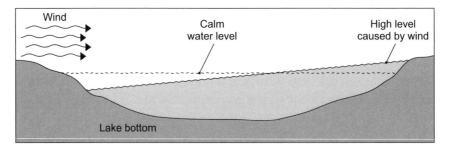

Figure 4.2 Initiation of seiche movement by wind; when wind ceases, water "sloshes" back to the other side of the lake, and back again in a slowly diminishing oscillation. *(Illustration by Jeff Dixon. Adapted from U.S. EPA Great Lakes Program 1995.)*

with a weather system crossing a large lake can have the effect of pushing down (or pulling up) on one end of the lake, setting up an up-and-down, back-and-forth pattern of water level known as a seiche movement. Winds can also create seiche movements, by pushing water via friction on the water surface to create not only waves but a general tide-like piling up of water on the downwind side of the lake (see Figure 4.2). Seiche movements create turbulence and can cause mixing at the edges of stratified lakes.

Life Zones in Lakes

The factors that influence the presence or absence of different organisms vary considerably within lakes: temperature, light availability, oxygen availability, nutrients, and food source. Three different life zones are distinguished, based on these and other factors. These life zones typically support distinct but interrelated communities of life within a lake.

The *benthic zone* is the lake bottom. The organisms that live in and on the bottom are known collectively as the benthos. The nearshore, shallow-water benthic zone is usually included in discussions of the littoral (shoreline) zone. Out beyond the littoral zone, benthic organisms are primarily microscopic decomposers (bacteria, fungi), detritivores (for example, insect larvae), and filter feeders (depending on whether there is sufficient light to support algae). In deeper water, the benthic community consists entirely of decomposers and detritivores, and the larger animals that prey on them. Substrates in the benthic zone are varied, but in any case a layer of mud covers whatever mineral substrate is there. In some lakes, this layer has a high organic content, in others not. It is subject to bioturbation, or mixing and physical disturbance caused by foraging of fish and movement of worms, molluscs, and other creatures.

The *littoral zone* is the nearshore area, sometimes defined as the area between low water levels and high water levels (corresponding to the intertidal zone or littoral in marine environments), and sometimes defined as the zone in which rooted emergent macrophytes (plants whose roots are in the lake bottom but whose leaves and flowers extend above the water surface) can exist. Some sources define it as the zone in which sufficient light penetrates to support submerged aquatic vegetation, but in shallow lakes, this definition loses the fundamental element of proximity to the shore.

Abundant light is one important characteristic of the littoral zone, since by definition it is a shallow-water habitat. Oxygen is abundant as well, both because of photosynthesis and (more important) because of turbulence and near-surface mixing. There are two types of littoral zones, based on the composition of the substrate: rocky shore littoral zones, and more sheltered shores characterized by sediment accumulation and emergent macrophytes.

The *pelagic zone* is the open-water zone; it is sufficiently far from the shores of the lake as to not be influenced by them. The pelagic zone is relatively featureless, consisting of water, air, wind, and sun. The upper levels are well oxygenated even under conditions of stratification because of diffusion of atmospheric oxygen, photosynthesis, and turbulent mixing. A surprising diversity of life is found in the pelagic zone, from phytoplankton and zooplankton, to small crustaceans like daphnia (members of genus *Daphnia*, order Cladocera), to large vertebrates like fish and seabirds.

The *abyssal zone*, sometimes called the profundal zone, occurs in deep lakes. Well below the influence of light, generally untouched by turbulence or mixing, but above the bottom layer, it is a dark, quiet, still environment. In deepwater environments, this deep region is considered part of the pelagic zone and in such cases is subdivided into layers based on depth and light penetration.

Lake Biota

In general, the types of plants and animals that populate lakes are similar to those that inhabit rivers and wetlands: aquatic plants, invertebrates, phyto- and zooplankton, fishes, waterfowl, and wading birds (see Chapter 1). It is impossible to specify a group of species that are "lake organisms," because there is so much overlap, and also because of the tremendous diversity of lakes: tropical lakes and arctic lakes would not be expected to share many species, for example, although some migratory bird species might be found on both in different seasons. However, distinct assemblages of organisms are found in different parts of the lake environment.

The benthic zone, away from the shoreline's turbulence, is typically covered with a layer of fine sediment rich with organic detritus. This mud provides both habitat and a source of food for detritivores such as midgefly larvae, nematodes and other worms, and molluscs such as snails and mussels. Because the benthic zone is often low in oxygen, particularly in deep lakes, its inhabitants have developed various adaptations for living in oxygen-depleted water. Benthic fishes, such as carp and catfish, rely more on taste and feel for finding food than on sight, since light is often low at the lake bottom. Macrophytes are usually absent from the benthic zone due to lack of light.

In the littoral, or inshore zone, the substrate may be either mud or coarser material such as gravel and cobble (rocks about the size of a fist). Except for shorelines exposed to high energy waves, there are likely to be macrophytes of various kinds. These provide a much more complex habitat for aquatic animals than the muddy bottom of the benthic zone, and there is a greater variety of such inhabitants. In the shallow littoral zone, emergent plants such as reeds and cattails along with floating-leaved plants like water lilies and pondweed are found. Because they are rooted in mud that is poorly oxygenated, emergent macrophytes typically have developed special structures to move air from the emergent leaves down to the roots. A little farther out from shore, in deeper water, completely submerged aquatic plants like coontail, watermilfoil, and water weeds are found.

Stands of littoral zone macrophytes are often nurseries for young fishes and other small animals, providing cover from predators. The littoral zone also is populated by insect larvae of various taxa such as dragonflies, mayflies, stoneflies, and caddisflies, as well as midge fly larvae. Some insects spend their adult lives in the littoral zone as well, including the waterboatman, water stick, and diving beetles. Fish, too, inhabit the littoral zone, feeding on the abundant invertebrate life there. Fish densities tend to be higher in the littoral zone than in either the benthic or pelagic zones. Some fish, like the predatory pike, live in the littoral zone; others breed there and spend their early life stages among the macrophytes. Amphibians and reptiles, particularly turtles, are found in the littoral zone. The abundance of invertebrates, fish, and amphibians attracts larger predators, including many wading birds like the green heron, which is found in lakes throughout the world.

Despite its austere appearance, the pelagic zone supports a diverse web of life, particularly in the photic zone. Phytoplankton, including diatoms, cyanobacteria, dinoflagellates, and algae, are multitudinous in eutrophic lakes. Herbivorous (plant-eating) planktonic animals feed on the phytoplankton and, in turn, are eaten by carnivores like the waterflea daphnia, and copepods, both crustaceans. The largest such zooplankters are less than 0.5 in (about 1 cm) long, most considerably smaller.

These zooplankters in turn are eaten by small fish. Some, like the alewife, consume zooplankton their entire lives; other pelagic fish do so when young but move on to larger prey as they grow. Typical pelagic fish include the salmonids (a family that includes salmon and trout) and the clupeids (the family of herring, sardines, and anchovies). Larger fish in the upper pelagic zone fall prey to grebes, ospreys, terns, and other birds.

Salt Lakes

Even though this volume's subject matter is the *freshwater* aquatic biome, a discussion of salt lakes is appropriate here because, although they are not freshwater, neither are they part of the marine biome. Salt lakes are a unique type of aquatic environment that is not easily categorized. Yet of the total water volume of the world's lakes, half is freshwater, and half is saline. The world's largest lake (although some call it a sea), the Caspian Sea, is saline. The highest lake in the world—Nan Tso, at about 16,000 ft (about 5,000 m) above sea level on the Tibetan plateau—and the lowest lake in the world—the Dead Sea, at about 1,312 ft (400 m) below sea level—are saline.

Saline lakes (see Table 4.3) are generally defined as lakes with a higher level of salts (ions of chlorine, sodium, magnesium, sulfur, calcium, and potassium) than freshwaters. Saline lakes are sometimes defined as lakes with salinity greater than 3 (using the Practical Salinity Scale). Ocean water on that scale is about 35, which corresponds to about 3.5 percent salt. But many saline lakes have much higher salinity than ocean water. The Great Salt Lake in Utah, for example, has a salinity that fluctuates between 110 and 330. Lake Asal, in Djibouti (eastern Africa), is the most saline lake, with average salinity of 350 (the Dead Sea sometimes reaches that level as a maximum).

Saline lakes have some physical characteristics that are similar to freshwater lakes, and some that distinguish them. Like freshwater lakes, they exhibit a range of mixing regimes—some are monomictic, some are polymictic, some stratify, some do not—depending in large part on the size, shape, and local climate conditions of the lake. Many saline lakes are relatively shallow, and shallow lakes tend not to stratify.

In those saline lakes that do stratify, and sometimes in those that do not, a halocline is often present. A halocline is a depth at which there is a relatively abrupt change in salt concentration, usually at the same depth as the thermocline. Some larger salt lakes are meromictic (having a stratified deep layer that never mixes) with very high salt concentrations in the hypolimnion and often very low oxygen as well.

Saline lakes resist freezing (salt water has a lower freezing point than freshwater). They tend to be alkaline and sometimes have quite high pH. Primary production is seldom nutrient limited; nutrient levels may be very high. Dissolved metals may be present, sometimes reaching concentrations lethal for some organisms.

Highly saline lakes are difficult living environments for organisms of almost every type. Typically, cellular organisms adapted to live in freshwater environments maintain salt concentrations in their cells that are higher than the surrounding water. This allows them to maintain cell turgor pressure, which, for example, keeps plants from wilting. Freshwater organisms face the problem of maintaining their elevated internal salt concentrations, since osmosis tends to equalize concentrations on both sides of a semipermeable membrane such as a cell wall or skin. The problem in effect is to prevent waterlogging, since the higher salt concentrations will draw water into the organism osmotically.

Organisms living in highly saline environments, on the other hand, face the opposite problem: drying out. This is an ironic fate in an aquatic environment, but if

Table 4.3 The World's Largest Saline Lakes

LAKE	COUNTRY	AREA [km²]	AREA [mi²]	VOLUME [miles³ (km³)]	MEAN DEPTH [feet (meters)]	MAX. DEPTH [feet (meters)]
Caspian	Russian Federation and Iran	386,400	149,200	18,713 (78,000)	614 (187)	3,363 (1,025)
Aral	Kazakhstan and Uzbekistan	68,000 pre-1960; 17,160 in 2004	26,300 pre-1960, 6625 in 2004	255 (1,064) originally	52 (16) originally	226 (69) originally
Balkhash	Kazakhstan	15,500–19,000	6,000–7,300	29 (122)	20 (6)	89 (27)
Eyre	Australia	9,300	3,700	6 (23)	10 (3)	20 (6)
Issyk-kul	Kyrgyzstan	6,280	2,425	417 (1,740)	902 (275)	2,303 (702)
Urmia	Iran	5,200–6,000	2,000–2,300	6 (25)	16 (5)	52 (16)
Qinghai or Tsing Hai	Tibet (China)	6,000	2,300	20 (85)	57 (17.5)	121 (37)
Great Salt Lake	United States	2,460–6,200	950–2,400	5 (19)	13 (4)	36 (11)
Van	Turkey	3,700	1,434	46 (191)	174 (53)	1,804 (550)
Dead Sea	Israel, Jordan	1,020	394	33 (136)	476 (145)	1,312 (400)

Sources: http://lakes.chebucto.org/saline1.html; and Encyclopedia Brittanica online.

lake salinity is higher than internal salinity, this is precisely what will happen. Halophiles (salt-loving organisms) have adapted to highly saline waters by following one of two strategies, both of which have to do with reducing the osmotic differential between their cells and the outside aquatic environment. In other words, they increase the concentration of dissolved materials inside their cells to a level nearer that outside. One strategy, used by a relatively small number of bacteria, involves using potassium ions. The other strategy involves increasing concentrations of certain dissolved organic materials. In either case, there is an energy cost to the organism. The benefit the organism receives in return is to be able to live in an extremely saline environment in which there is relatively little competition for resources. For larger organisms, such as crustaceans, there is often little predation from fishes, although bird predation may be very high.

Plants have similar adaptations on the cellular level but also have additional adaptations, including organs whose purpose is to keep salts from entering. In animals, in addition to the cellular adaptations described above, organ systems are developed that collect and excrete salts. Organisms that maintain a preferred internal osmotic concentration by excreting salts are called osmoregulators, and those that match (within limits) the external osmotic environment by building up their internal osmotic concentration are called osmoconformers.

Many salt lakes are ephemeral; they appear during wet periods and then dry up. Or, if not ephemeral, they vary greatly in size and volume over the course of the year. Salt lakes are all endorheic, which means that they do not have outlets. Water, when it is flowing or precipitating, enters them, but leaves only by evaporation. When water evaporates, it leaves behind its dissolved salts, and over time these salts accumulate in the basin. Some salt lakes, such as the Great Salt Lake in Utah, are the remnants of what were once much larger lakes, which have slowly evaporated as climate changed. In some cases, the shrinking of lakes into highly saline, smaller lakes has been caused by human interventions, intentional or unintentional. The Aral Sea in Central Asia, and Mono Lake in California, are examples. In each case, water from the lake's tributaries was diverted. In the case of Mono Lake, it was to supply water for the Los Angeles region. In the case of the Aral Sea, it was to supply water for irrigation agriculture in the former Soviet Union.

High salinity is not the only problem faced by organisms in saline lakes, or even the greatest problem. Lakes that fluctuate widely and frequently in volume also fluctuate widely and frequently in salinity, so organisms living in them must be able to tolerate changing salinity levels, sometimes over orders of magnitude. As noted, some salt lakes dry up annually, so these organisms must be able to survive periods of increasing salinity followed by complete drying out (dessication). This eliminates fish, although in less extreme conditions the lungfish can survive by breathing air, burrowing down into the mud, and estivating (like hibernating except in the summer, not the winter). Smaller organisms adapted to extreme environments of high and variable salinity along with dessication typically employ a strategy of encysting, which is to say encapsulating themselves, usually but not always at the egg life stage.

Dissolved oxygen levels typically decrease as salinity increases. Low oxygen is usually a fact of life in salt lakes. Another influence on oxygen levels is temperature, and in arid environments where salt lakes are found, the dry season is often a hot season as well. The warmer the water temperature, the lower the oxygen levels, independent of salinity.

Lakes with extremely high salinity, like the Dead Sea, have only a few species living in them. As a rule, the higher the salinity of a lake, the fewer different species will be found in it, and the simpler the food web will be.

Saline lakes vary considerably in the relative proportions of different elements that contribute to their salinity. Some ions, such as sulfate, create much more difficult environments for living organisms than others. In such environments, only highly specialized, usually microscopic, organisms can survive.

Which types of bacteria, phytoplankton, and zooplankton dominate depend on specific lake conditions, particularly the relative proportion of different salts. In lakes with varying salinity, microorganisms that are abundant one year may be barely detectable the following year. Green sulfur bacteria have been found in lakes with extremes of salinity and high levels of sulfides. In the Dead Sea, green algae and red archaeobacteria are seen in great concentrations in years when salinity levels decrease. Species of these two groups of organisms are found in practically every saline lake. The red archaeobacteria are responsible for the red or pink coloration sometimes seen in salt lakes during "blooms" under optimal conditions.

Zooplankton are also highly sensitive to salinity levels, although some species have broad tolerances. Rotifers did well in Mono Lake at salinities of 60–70 ppt, but when the lake exceeded 80 ppt (as it continued to shrink during a period when most tributary water was diverted), the rotifers were no longer detectable. Lake Eyre in Australia, whose salinity ranged from 25 ppt to 273 ppt over the course of a single year (1985), is home to several species of rotifers, crustaceans, cladocerans, and copepods.

What is true for microorganisms—species diversity decreases with increasing levels of salinity—is also true for macrophytes (nonmicroscopic plants) in salt lakes. Sago pondweed, widgeongrass, and spiral ditchgrass are among the very few vascular plants that can tolerate hypersaline waters (defined as greater than 50 ppt, or 5 percent salt). Only five species of emergent plants—the cosmopolitan bulrush, desert saltgrass, Nuttall's alkali grass, chairmaker's bulrush, and seaside arrowgrass—occur regularly over a range of saline lakes including the hypersaline.

Brine shrimp are important primary consumers in many salt lake food webs and are one of the few macroinvertebrates that can live in hypersaline environments. Brine shrimps are found in nearly every large salt lake in the world, often in very large numbers that attract vast flocks of waterbirds. Figure 4.3 illustrates the prominent role of brine shrimp in both benthic and pelagic food webs of Great Salt

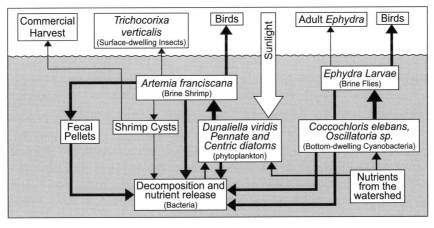

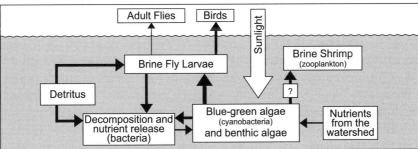

Figure 4.3 Trophic diagrams for Great Salt Lake, Utah: (top) planktonic (that is, pelagic) habitat and (bottom) benthic habitat. *(Illustration by Jeff Dixon. Adapted from USGS Utah Water Science Center 2001.)*

Lake, Utah. Other, larger crustaceans are present in some salt lakes. In Lake Eyre, Australia, inhabitants include a crayfish (the "yabbie") and an atyid prawn.

Salt lakes are often home to large insect populations, though typically of relatively few species. At Mono Lake, the alkali fly is abundant. Some midges tolerate salt well and are found in salt lakes, sometimes in great abundance. Several species of damselflies and dragonflies are also halophiles.

Fishes are present in most saline lakes, even hypersaline lakes, where they survive in refugia (refuges where conditions are more suitable) such as the estuaries of freshwater tributaries or springs. Lake Eyre contains the desert goby, the dalhousie hardyhead, an endemic hardyhead, the western mosquitofish, and a catfish. Some tilapia also can live in moderately saline waters.

While saline lakes have relatively few species, the ones that are there often reach astronomical populations, sometimes seasonally. Huge populations of zooplankters, brine shrimp, and insect larvae make such lakes attractive to waterbirds, both resident and migratory. African saline lakes support at times several million flamingos and other wading birds. At Mono Lake in California, more than 1.5

million eared grebes typically visit during the fall migration, in addition to about 100 other species.

Manmade Lakes

People have been building dams to hold back and tame the waters of rivers since the dawn of civilization; indeed some people have argued that the building of large-scale water development projects (dams and canals, irrigation ditches) was inextricably bound up with the rise of civilization. The first large dam was built in Egypt nearly 5,000 years ago; 46 ft (14 m) high and 371 ft (113 m) wide at the top, it was apparently built as a flood-control dam. It seems to have been destroyed by a flood before it was complete, so it is unlikely that it created any standing body of water, at least not for long. Dam-building accelerated during the Roman empire, but it was not until the middle of the twentieth century that the building of large dams and the creation of large lakes grew explosively. By the beginning of the current century, there were at least 45,000 large dams in the world, and half of the world's rivers have at least one dam on them. Dam-building has slowed considerably in recent decades in the industrial countries for two reasons: the influence of the environmental movement has increased awareness of the environmental and social costs of large dams, and the fact that most of the suitable sites for large dams already have large dams on them. In developing countries, the building of large dams has accelerated; a milestone of sorts was reached when China recently completed the world's largest dam project, the Three Gorges Dam on the Yangtse (or Chang Jiang) River.

Dams are not the only way that lakes are created by human activities—for example, abandoned quarries often fill with water—but they are by far the most important, so this section focuses on lakes created by the building of dams. Three main differences distinguish impounded waters in reservoirs from natural lakes.

First, the influence of the watershed is likely to be greater in manmade lakes, because the size of the watershed in relation to the size of the lake is usually greater. For example, Lake Powell, created by a dam on the Colorado River in Arizona, holds at most 6.5 mi^3 (27 km^3) of water in a watershed of 102,812 mi^2 (266,283 km^2), while Lake Superior in North America has 2,735 mi^3 (11,400 km^3) of water in a watershed of 49,305 mi^2 (127,700 km^2). The ratio of the area of watershed to the amount of water stored is about 1,000 times greater for Lake Powell.

Rivers carry more than just water off their watersheds, and the larger the watershed, the more sediment, organic matter, nutrients, and pollutants are carried into impoundments. Though water passes fairly rapidly through impoundments, these other materials are subject to different fates. The most obvious of these materials, and perhaps most troublesome to the operators of dams, is sediment. Sediment particles—everything from clay, silt, and sand to gravel, cobble, and boulders—are carried by rivers, moved by the energy of flowing water. When flowing water hits the backwaters of a large impoundment, this energy is no longer available and sediments sink to the bottom. Deltas may form where tributaries enter the reservoir; given enough time, sediment will fill a reservoir, impairing the function for which it was constructed.

Along with sediment, organic material is trapped. In reservoirs subject to stratification, this organic material may contribute to the development of low-oxygen conditions in the hypolimnion as it decomposes. Many hydroelectric dams release water from relatively deep within the reservoir, and in many cases, the release of hypoxic (low oxygen) water has caused environmental damage downstream. Not all reservoirs are subject to stratification, however; shallow reservoirs and reservoirs with very low retention time are less likely to experience stratification.

The anaerobic decomposition of organic material in reservoirs is also a concern because it may be a significant source of methane in the atmosphere. Methane is a potent greenhouse gas, and its increasing concentration in the atmosphere is thought to contribute to global climate change. Methane releases from reservoirs are of particular concern in the Tropics, where large areas of tropical forest have been flooded as reservoirs filled. Decomposition is rapid in the warm tropical waters, and methane releases have been substantial.

The availability and biochemical transformations of nutrients are affected by reservoir conditions. This may be as much a problem downstream as it is for the reservoir ecosystem. Biological activity in reservoirs can result in different retention times for nutrients than for water, and the effects may be different for different nutrients. For example, if a reservoir selectively traps a particular nutrient but allows others to pass through, downstream ecological impacts may be observed. Recent studies suggest that a cascade of effects is resulting from Iron Gate Reservoir on the Danube River, a tributary of the Caspian Sea. The reservoir is acting as a silica trap; silica concentrations in the river downstream are significantly reduced. This reduction in turn is being blamed for a shift in the composition of the phytoplankton community away from diatoms (which require relatively large amounts of silica) and toward blue-green algae. This shift is reportedly harming fishing in the Caspian Sea.

Second, the average residence time of water is considerably shorter in manmade lakes than natural ones; this means that the throughput of water is greater. The difference in watershed area per volume makes this difference unsurprising: a larger area of land drains into a manmade lake than into a natural lake of equivalent size, so one would expect more water to come into (and hence go out of) manmade lakes. Reservoirs are, after all, dammed-up rivers. The residence time of water in Lake Powell averages 7.2 years, whereas in Lake Superior, it is close to 200 years. The median retention time for large reservoirs in the United States is little more than 100 days.

Third, reservoirs are subject not only to natural phenomena (droughts, cold spells) but also are subject to human manipulation of water levels, to say nothing of intensive management of lake resources such as fisheries. Dams are built for a variety of purposes: hydroelectric power generation, flood control, storage of seasonal high water for irrigation and city water supply, and low flow augmentation, as well as combinations of these purposes. The operation of a dam directly affects the conditions in the reservoir, and operations are different depending on the purpose of the dam.

The ability of a flood-control dam to control floods is dependent on having plenty of storage capacity available when the flood comes. The operations of such a dam can lead to dramatic changes in lake depth over the course of a year. Less dramatic, but more frequent, are the ups and downs of the lake surface as a result of hydroelectric power generation. Hydroelectric power dams typically operate to provide peaking power, to match daily peaks in demand for electricity. Operations usually involve large water releases for part of a day and small water releases for the rest of the day, during which time reservoir levels return to the prerelease level. The ups and downs occur on a weekly basis in some cases. Depending on the size of the reservoir and the generating capacity of the dam, these fluctuations in the lake surface elevation can be several feet or more daily, creating difficult conditions for organisms in the littoral zone, as well as causing bank erosion.

The loading of sediments, nutrients, and organic material into reservoirs is much faster then the loading in natural lakes. This rapid loading, in combination with generally unstable biological populations and food webs, has led to a number of observed environmental problems. Sediment accumulation and development of low-oxygen conditions in the hypolimnion occur, with methane generation as a secondary effect of decomposition under low-oxygen conditions. Another secondary effect of anaerobic decomposition in some tropical lakes has been the production of hydrogen sulfide in the hypolimnion. This toxic, flammable gas is then released in the dam tailrace as the pressure on the water is relaxed. The same biochemical transformations that produce hydrogen sulfide also acidify water in the hypolimnion, and this can damage the turbines in hydroelectric dams.

Some reservoirs become eutrophic, with blooms (population explosions) of phytoplankton resulting. With unstable biotic communities and plenty of nutrients, some newer reservoirs are susceptible to invasion by aquatic nuisance plants such as the floating water hyacinth; thick blankets of these plants have covered the surfaces of reservoirs in Africa and elsewhere, blocking out light and causing extreme damage to reservoir fisheries and recreational uses. Dams and reservoirs also have significant impacts on downstream river segments (see Chapter 2).

In the next sections, three natural lakes are examined in some detail. They were chosen to represent lakes at low-, mid-, and high-latitude locations, as well as to represent a variety of conditions.

An Ancient, Boreal, Deep, Relatively Pristine Oligotrophic Lake: Lake Baikal

Characteristics of the Lake and Its Watershed

Lake Baikal is the deepest lake in the world, at 5,370 ft (1637 m); it is the oldest freshwater lake in the world, at approximately 25 million years old; and it is the largest lake by volume, holding 5,662 mi^3 (23,600 km^3) of freshwater, 20 percent of the world's unfrozen total. This rift valley lake is long and narrow, almost 400 mi (636 km) long and 17–50 mi (27–80 km) wide. Its surface elevation is 1,496 ft (456 m) above sea level. Average depth is 2,499 ft (730 m), but depth varies greatly from place to place. The lake's surface area is 12,200 mi^2 (31,600 km^2), and the length of its coastline is nearly 1,300 mi (2,100 km).

Lake Baikal has three deep zones separated by ridges. The deepest, down 5,250 ft (1,600 m), is in the middle. Although a maximum water depth of nearly a mile is impressive, sediments have been collecting in the depths of the lake for millions of years and are thought to be more than 5 mi (8,000 m) deep.

Lake Baikal is remote, lying in Siberian Russia near the Mongolian border; part of the watershed is in Mongolia (see Figure 4.4). At 227,800 mi^2 (590,000 km^2), Lake Baikal's watershed is considerably larger than the State of California in the United States and straddles the border between Russian Siberia and Mongolia. It is mostly underlain by metamorphic and igneous rock, which accounts for the low levels of total dissolved solids in the tributaries and the lake itself. The watershed is mostly covered with boreal forest and steppe.

Lake Baikal has well over 300 perennial tributaries. The largest is the Salenga River, the source of about 50 percent of the water, entering the lake on the east side about a third of the distance from the southern end. It has created a huge delta, 20 mi (32 km) wide and around 100 mi^2 (258 km^2) in area. The Barguzin, the

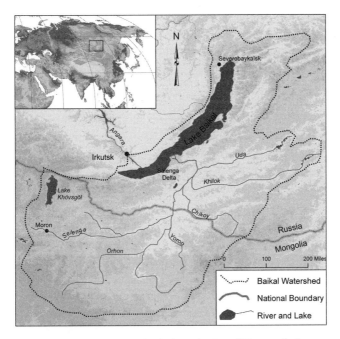

Figure 4.4 Lake Baikal and its watershed. *(Map by Bernd Kuennecke.)*

Upper Angara, the Bolshaya, the Kabanya, the Tompuda, the Tyya, the Goloust-
naya, the Vydrinaya, and the Snezhnaya are all major tributaries. The Angara
River is the lake's sole outlet, flowing north from the western end of the lake with
an average discharge of 67,000 ft³/sec (1,902 m³/sec). The Angara is a tributary of
the Yenisei, which flows into the Arctic Ocean.

Conditions in the lake. Baikal is an oligotrophic lake with the startling clarity that
usually accompanies low nutrient content. Its Secchi depths are up to 131 ft (40 m)
in the pelagic zone, even in June when phytoplankton are abundant. Phytoplankton levels are relatively low due to the low nutrient concentrations, and levels of
dissolved and suspended solids are exceptionally low as well. Some loss of clarity
is attributed to dissolved organic materials originating in marshes along the tributaries. Levels of total dissolved solids are low in the tributaries (81–134 mg/L among
the major tributaries) and in the lake itself (100–120 mg/L).

While many deep lakes, particularly in the tropics, stratify with resulting low-
oxygen waters in the hypolimnion, conditions in Baikal result in relatively
homogeneous conditions throughout the water column. Dissolved oxygen is high
(9 mg/L) even in the abyssal zone. One consequence of this fact is a well-
developed fauna, with many endemic species inhabiting the abyssal zone. Dissolved mineral levels are remarkably homogeneous with depth, but nutrient levels
show more variation. In general, the lake is more oligotrophic in the northern half
and less oligotrophic (though hardly eutrophic) in the southern half.

The lake is cold. The great bulk of the water rarely gets above 39° F (4° C); in
the deep abysses, it is closer to 37° F (about 3° C). The exception to the year-round
cold temperatures occurs in relatively shallow waters (the top 800–1,000 ft or 240–
300 m), and in the shallow bays around the lake perimeter. At the surface, water
temperature can reach 57°–64° F (14°–18° C) in later summer. Stratification occurs
with a sharp thermocline and, by 800–1,000 ft (240–300 m) depth, cooling to
between 37° and 39° F (3° and 4° C), temperatures that continue to the bottom.
The lake is also stratified in winter, with the coldest water temperatures reached in
the surface ice, and warmer waters below. There are two periods, one in June and
one in September, when a consistent temperature in the pelagic zone extends from
the surface to about 800–1,000 ft (240–300 m), allowing regular mixing to occur.
Below that depth, the water temperature is stable and only limited mixing occurs.

The Lake Ecosystem
Lake Baikal has been of great interest to scientists for decades because of its rich diversity of life and the relatively high proportion of species unique (endemic) to
Lake Baikal. There are 2,565 known species of animals, a third of which are macroinvertebrates and 80 percent of which are endemic. About 1,000 species of plants
are present, of which about a third are endemic. Baikal's isolation and long continuous existence have set the stage for a great deal of speciation among some groups
of plants and animals.

Plants. Photosynthesis takes place down to depths of more than 500 ft (~150 m) in Lake Baikal, giving it an extraordinarily large volume for phytoplanktonic activity. The lake is strongly autotrophic, with phytoplankton activity accounting for 90 percent of the organic input into the lake annually. The approximately 300 species and subspecies of algae belong to 111 genera, 56 families, and eight phyla. Green algae species are most diverse with 112 taxa, followed by cyanobacteria (blue-green algae), diatoms, and chrysophytes. The littoral zone is home to many benthic algae, but there are no macrophytes in this zone.

Phytoplankton production of biomass amounts to 3,925,000 tons per year. A large but annually varying proportion of Baikal's photosynthetic biomass production occurs in late winter and early spring while the lake is still covered with ice. The diatom group Melosira, as well as peridinean species, were dominant in different years in surveys of algae below the ice. Many of the algae are endemic.

Baikal has a number of hydrothermal vents, around which have developed communities of life adapted to the high temperatures. Photos of the vents reveal a white filamentous bacterial mat, along with sponges, amphipods, gastropods, and other unidentified organisms. It is not known whether the observed thermal vent community consists of vent-specific organisms or representatives of the normal benthic biota concentrated around the vents.

Invertebrates. Invertebrates in Lake Baikal are grouped according to their broad habitat: benthic, inhabiting the surface and subsurface sediments of the lake bottom; benthic-pelagic, living near but not on the bottom; and pelagic. The benthic invertebrates make up the largest group, with more than 1,000 species.

One of the prominent forms of life in Lake Baikal is freshwater sponges, which have speciated in Baikal to an extraordinary degree. Sponges look more like plants than animals; they have no central nervous system and are sessile (they cannot move). Some of Baikal's sponges are large, reaching several feet (about 1 m) in height. They cover almost half of the surface area of the lake bottom from about 6 to 130 ft (2 to 40 m) depth, and grow in such profusion that they resemble an underwater forest. Their green coloring is due to zoochlorella, algae that live symbiotically on and in the sponges. The dominant sponge is one of many endemics in the family Lubomirskiidae. Many of the sponges are colonized by an amphipod crustacean. Baikal's sponges are such effective filter feeders that they create a zone in which picoplankton are depleted. Picoplankton are very small (0.2–2 μm in diameter) algae and cyanobacteria. The majority of Baikal's sponges are endemic demosponges (a group of sponges whose structure is made up primarily of spongin, silica, or a combination).

Another invertebrate group that has diversified to an unusual degree in Baikal is the arthropods, particularly the gammarids (a type of amphipod). Freshwater amphipods are known in North America as scuds. In Lake Baikal, 265 species and 81 subspecies are classified into 51 genera. All but one are endemic. Gammarids reach impressive population densities, up to 2,800 individuals per square foot of substrate in the littoral zone. Similar species diversification has been observed in Lake Baikal among the oligochaete worms, midge flies, ostracods, and molluscs.

A single arthropod species, however, is the dominant pelagic species and extremely important in the food web. Although only 0.06 in (1.5 mm) long, the copepod epischura inhabits the entire water column, and its population reaches enormous numbers. It is estimated that the lake's epischura population filters annually between 120 and 240 mi^3 (500 and 1,000 km^3) of water, making it a major ecosystem engineer in the lake, contributing to the lake's clarity. It is the foremost consumer of phytoplankton in the lake. Epischura is a major prey species for one of Baikal's famous fish, the anadromous salmonid Baikal omul, and is thus a critical link between production in the lake and higher organisms.

Gigantism (very large size of individual specimens) is common among Lake Baikal's animals; not only many invertebrates but even some phyto- and zooplankton species are exceptionally large. The amphipod fauna of Baikal is characterized by gigantism (individual gammarids are exceptionally large) and the world's largest known amphipods are Baikal endemics. This gigantism is attributed to the very high concentration of oxygen in the lake. The world's largest planarium (flatworm), *Baikaloplana valida*, which grows to 16 in (40 cm) in length, lives in Baikal along with several other giant flatworm species. The molluscs of Baikal also tend toward gigantism.

Fishes. As is the case with other taxa, Baikal is known for unique and endemic fishes. Baikal's unusually high-oxygen concentrations even at great depth have created conditions for the evolution of an extraordinarily large and diverse community of abyssal fishes. Fifty-two of the 58 species and subspecies of fishes in Baikal are native; the remainder are introduced.

Fishes are of four groups: Eurasian native species, Siberian native species, Baikal endemic species (see Table 4.4), and introduced (nonnative) species. The native species have in effect partitioned the habitat. Fishes of Eurasian taxa inhabit the protected shallow bays (*sors*) and river deltas; Siberian fishes inhabit the littoral zone and outfalls of rocky mountain streams; and endemic species inhabit the deep littoral to abyssal zone. Table 4.5 shows the distribution of species by family in various categories.

Baikal's famous fishes include the Baikal omul, a salmon-like migratory fish that is economically important to the Baikal region. The transparent golomyanka are one of the most populous vertebrates in Baikal and a principal food source for the Baikal seals (nerpas).

Mammals. The best-known mammal in Lake Baikal is a freshwater seal, the endemic Baikal nerpa. It is Baikal's only endemic mammal. In contrast to the many examples of gigantism among Baikal's animals, the nerpa is one of the smallest seals. Adults reach perhaps 4.3 ft (1.3 m) in length and 139–154 lb (63–70 kg); males may be larger than females but the difference is slight.

Table 4.4 Endemic Fishes of Lake Baikal

Family	Species
Abyssocottidae (deep water sculpins)	*Asprocottus herzensteini* Erg
	A. intermedius Taliev
	A. platycephalus Taliev
	A. abyssalis Taliev
	A. pulcher Taliev
	A. parmipherus Taliev
	Limnocottus godlewskii
	L. megalops Gratzianow
	L. eurystomus Taliev
	L. griseus Taliev
	L. pallidus Taliev
	L. bergianus Taliev
	Abyssocottus korotneffi Berg
	A. gibbosus Berg
	A. elochini Taliev
	Cottinella boulengeri Berg
	Procottus jeittelesii Dyb.
	P. major Taliev
	P. gurwici Taliev
	Neocottus werestschagini Taliev
Acipenseridae (sturgeons)	*Acipenser baeri baicalensis* Nikolsky
Comephoridae (Baikal oilfishes)	*Comephorus baicalensis* Pallas
	C. dybowski Korotneff
Coregonidae	*Coregonus autumnalis migratorius* Georgi
	C. lavaretus baicalensis Dyb.
	C. lavaretus pidschian Gmelin
Cottidae (sculpins)	*Cottus kesslerii* Dyb.
	Paracottus iknerii Dyb.
	Batrachocottus baicalensis Dyb.
	B. multiradiatus Berg
	B. talievi Sideleva
	B. nikolskii Berg
	Cottocomephorus grewingkii Dyb.
	C. inermis Jakowlew
	C. alexandrae Taliev
Thymallidae	*Thymallus arcticus baicalensis* Dyb.
	T. arcticus brevipinnis Svetovid

Source: Sideleva 2000.

Nerpas spend most of their life in the water. In winter, they maintain breathing holes but do not climb on to the ice until spring, when they give birth. Mothers dig large dens for their pups in snow drifts or ice hummocks, and nurse them longer than any other seal. In late spring when the ice and dens melt, the pups are left to

Table 4.5 Fish Taxa of Lake Baikal

Family (number of species/subspecies)	Endemics	Native but Not Endemic	Introduced	Shallow Water (including bays/ deltas)	Deep Water	Deep and Shallow
Abyssocottidae (20)	20			3	16	1
Acipenseridae (1)	1			1		
Balitoridae (1)		1		1		
Cobitidae (1)		1		1		
Comephoridae (2)	2				2	
Coregonidae (5)	3		2	4		1
Cottidae (9)	9			3	3	3
Cyprinidae (10)		8	2	10		
Eleotridae (1)			1	1		
Esocidae (1)		1		1		
Lotidae (1)		1		1		
Percidae (1)		1		1		
Salmonidae (2)		2		2		
Siluridae (1)			1	1		
Thymallidae (2)	2			2		
Total (58)	37	15	6	32	21	5

Source: Sideleva 2000.

fend for themselves. Mating among adults begins fairly soon after mothers have given birth. After the ice melts, nerpas tend to occupy the littoral zone and shoreline.

Nerpas feed on fish, primarily the golomyanka, which they hunt underwater. They are able to hold their breath for up to an hour at a time although their feeding forays usually last only 10 to 20 minutes. Nerpas also eat invertebrates. They have been observed foraging to depths of 330 ft (100 m).

Besides the famous nerpa, the only other aquatic or semiaquatic mammal existing in Lake Baikal is the otter. Otters inhabit the tributary rivers and littoral zone and feed on fish, molluscs, and crustaceans. Terrestrial mammals on the north side of the Lake are typical of taiga fauna, with 39 species including brown bear, wolverine, and the highly prized sable; on the southern side, there are 37 mammal species.

Birds. Lake Baikal is on two major migratory bird flyways: the Selenga and the Khingan. An estimated 10–12 million waterfowl pass through the Baikal region during the southward (fall) migration. Waterfowl, including dabbling ducks, diving ducks, geese, and swans, first begin to arrive in the lake region in late March, and the spring migration continues into May. Migrants include Mallards, the Common

Teal, Northern Pintail, Eurasian Wigeon, and Common Goldeneye. A number of rare species, such as Bewick's Swan, use the region.

Problems and Prospect

Until quite recently, Lake Baikal was considered to be as pristine a large lake as could be found. Its relatively pristine state was not surprising considering the remoteness of its location and the low population density and land use intensity of its watershed. The perception of Lake Baikal has changed recently, however, as more studies are completed on the lake and its biota, and as evidence of the effects of human activities accumulates.

Land use in the watershed is of low intensity, consisting primarily of forestry and grazing. Mineral and energy extraction is becoming more important. Population density is low on average, but higher around the southern end of the lake. Several significant industrial centers exist in tributary watersheds on the south end of Lake Baikal. A large pulp and paper mill, the Baikalsk Pulp and Paper Mill, is located on the southern shore of the lake and is the only large industrial facility on the lake.

During the Soviet era, a series of hydropower dams and reservoirs was constructed on the Angara River downstream of the lake. The uppermost, Irkutsk Reservoir, was constructed so that its backwaters would back up into Lake Baikal. The reservoir's maximum water surface elevation is high enough that project operators can, in effect, use Lake Baikal as an auxiliary water storage basin. However, because of the size of the lake, the operations of the Irkutsk project do not cause frequent fluctuations of water level in Lake Baikal. Over a period of 50 years, however, the lake's surface elevation has risen about 3 ft (1 m) overall. Environmental damage attributed wholly or in part to this raising of the water level includes shoreline erosion and degradation of fisheries.

Water quality in general is better on the north end of the lake; the south end receives some pollution from the large pulp mill on the lake shore as well as nonpoint source pollution from the Salenga River.

Pollution is apparently taking its toll on the lake's web of life. The nerpa population is declining rapidly as the animals' health deteriorates and death rates increase. The nerpa population was last officially counted in 1994, at which time there were 104,000. Six years later, a scientific expedition estimated that there were only 85,000 and that the death rate among nerpa had tripled. An environmental organization, Greenpeace, performed their own count and estimated no more than 65,000 in 2001. By all accounts, the population is declining rapidly and the spread of diseases has been observed. Bioconcentration and biomagnification of organochlorines, including dioxin, in the food web are suspected to be the culprit in this case.

The nerpa population decline is also attributed to another factor besides pollution: hunting. Nerpa pups are hunted for their valuable pelts; adults are hunted for sport. There are legal quotas, but the rate of illegal hunting is thought to be high due to inadequate law enforcement in the region.

The general health of fish and other organisms has deteriorated, and mutations are observed in commercially valuable fishes from parts of the lake receiving industrial discharges. The bioconcentration and biomagnification of organochlorine chemicals including dichlor-diphenyl-trichloroethane (DDT) and polychlorinated biphenyls (PCBs) are taking place in organisms from a variety of taxa in the lake's food web. Body concentrations, though worrisome, were found to be an order of magnitude lower in Baikal Herring Gulls than in Herring Gulls from the highly polluted Lake Ontario. It is thought that the pesticides detected in migratory birds (who also happen to be top predators when they are present at the lake) come from agricultural and mosquito control practices in southern Asia, rather than the waters or food chain of Baikal. However, even resident birds of Lake Baikal carry considerable pollutant burdens in their bodies.

Pollutants in the lake and its tributaries are blamed for observed declines in some fishes including the omul. Changes in the planktonic community have been observed over the past five decades, with endemic species declining and being replaced by more widespread (cosmopolitan), pollution-tolerant species. The 2006 decision by the Russian government to build a controversial oil pipeline traversing the watershed of Lake Baikal will create the possibility of a large-scale oil spill. Infrastructure is also being planned to "open up" remote parts of the Mongolian section of the watershed for Chinese-sponsored mineral and energy development. Ultimately, such developments may spell the end of Lake Baikal's pristine state. Relatively undeveloped natural resources are in great demand. As the world's economies grow, the riches of Lake Baikal and its region may simply be too great a temptation.

Increasing international attention is being focused on the conservation needs of the lake. The Selenga delta was designated as a wetlands of international significance under the Ramsar Convention. Lake Baikal was added to the list of World Heritage sites maintained by the United Nations Educational, Scientific and Cultural Organization (UNESCO). This designation draws attention to the global significance of the lake and lends support to those working to preserve the lake's environment. Perhaps in part as a result of UNESCO-sponsored studies and publicity, several national and international environmental organizations have mounted campaigns to reverse the decline of Lake Baikal. It remains to be seen whether sufficient political will and resources can be mustered in the region to take the actions necessary to reduce pollution and protect the lake and its rich biodiversity.

A Tropical, Shallow, Eutrophic Lake Heavily Affected by Human Activities: Lake Victoria

Lake Victoria is one of the East African Great Lakes, along with Lake Tanganyika and Lake Malawi. Lake Victoria is near the others yet very different. Tanganyika and Malawi are both rift valley lakes, like Lake Baikal. And like Baikal, they are

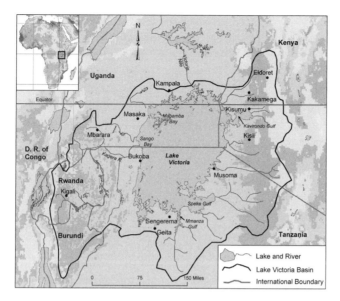

Figure 4.5 Lake Victoria and its watershed. *(Map by Bernd Kuennecke. Adapted from Kayombo and Jorgensen 2006.)*

long, deep, narrow lakes. Lake Victoria, by contrast, is relatively shallow and, while more rectangular than round, it is anything but narrow (see Figure 4.5). It also contrasts with those other lakes in that it is much younger, about 800,000 years, although its continuous existence over such a long time is in doubt. There is some evidence that it dried up completely 12,000 to 15,000 years ago.

Lake Victoria is a lake of almost mythical productivity, supporting millions of people with its fish. But it is also a lake undergoing rapid change, and its future is in doubt.

Characteristics of the Lake and Its Watershed

Lake Victoria is the largest lake in Africa (by surface area) and is second in the world; only Lake Superior in North America is larger. It is however a relatively shallow lake, and its watershed is small compared with the size of the lake. Three countries control the lake: Tanzania (49%), Uganda (45%), and Kenya (6%). The divided jurisdiction gives rise to management challenges.

The lake is relatively young, although how young is the subject of considerable debate. Many include it among the so-called ancient lakes of the world, but if the criterion is *continuous* existence for 100,000 years, Lake Victoria may not qualify. It is certainly younger than its fellow East African Great Lakes, but they are ancient indeed. Lake Victoria's origins lie in the uplift of a large region of East Africa, which in the process created a slightly concave land surface.

The watershed lies entirely in the tropical zone—the Equator actually crosses the lake. The northern lakeshore area and Ugandan section of the basin are for the

Lake Victoria by the Numbers

Lake coordinates: 0° 20′ N–3° 0′ S, 31° 39′
    E–34° 53′ E

Elevation: 3,720 ft (1,134 m)

Surface area: 26,563 mi^2 (68,800 km^2)

Watershed area: 74,517 mi^2 (193,000 km^2)

Residence time (volume/outflow): 140 years

Maximum depth: 262 ft (80 m)

Mean depth: 131 ft (40 m)

Volume: 2760 km^3

Shoreline length: 2,137 mi (3,440 km) (not
    including islands)

Surface temperature: 75°–86° F, 24°–30° C
    (approximately)

Bottom temperature: 73°–81° F, 23°–27 C
    (approximately)

most part classified as equatorial fully humid or equatorial monsoonal (Koeppen climate classes Af and Am, respectively). Equatorial or tropical humid regions are typical of tropical rain forests, with no real seasons and at least 2.4 in (6.1 cm) of rain every month, while in monsoonal regions, winds reverse directions according to the seasons, and there is typically at least one relatively dry month. The southern half of the basin is primarily in the tropical wet and dry climate zone (Koeppen Aw). Tropical wet and dry climates are associated with the savanna biome and are characterized by pronounced seasonality of precipitation, often with a long, very dry season.

As one would expect from an equatorial region, temperature is uniformly warm, with little seasonality and little difference from one part of the basin to another. Annual average minimum temperatures are typically in the range of 60°–65° F (15.5°–18.5° C); annual average high temperatures range from 81°–88° F (27°–31° C), depending on location.

The dry season in the case of Lake Victoria and its basin is generally in June through August, and the southeast quadrant of the basin is the driest. Even in the northwestern part of the basin, which is the wettest, the period June through October is usually the driest. March through May is the wettest period throughout the basin, but there is another wet period in November and December. This latter wet period is known as the Little Rains (as distinct from the Big Rains) and is not completely reliable.

Consistent with its large surface and relatively small volume, precipitation falling on the lake accounts for the bulk of water coming into it (more than 80 percent) and evaporation and transpiration together account for most of the water going out of it. All told, more than 260,000 ac-ft (320 million m^3) of water fall on the lake and its watershed in the form of rain (1946–1970 average). But, due to its equatorial location and climatic conditions, evaporation and transpiration losses from both the lake and the watershed amount to 235,000 ac-ft (290 million m^3). If the lake level stays constant, close to 25,000 ac-ft (30 million m^3) of water leaves the lake each year as surface flow. This output is via the Victoria Nile River, a tributary of the famous Nile River, at the northern end of the lake.

The Kagera River, flowing into Lake Victoria from the west, is the largest tributary and carries more than half of the surface water inflow to the lake. Another 10 relatively small tributaries each add about 4 percent on average.

The lake basin is roughly saucer-shaped. Its shoreline is rocky, intricately indented, and changes considerably with fluctuations in water level. There are several large bays (gulfs), including Mwanza Gulf, Speke Gulf, and Kavirondo Gulf, as well as countless small bays. The numerous islands large and small give the lake an extensive littoral zone. Fringe marshes dominated by papyrus are common in the protected bays.

Biota of Lake Victoria

Plants. Lake Victoria's phytoplankton is diverse. More than 600 species are reported, belonging to 117 genera. This flora is composed of cyanobacteria, green algae, and diatoms, although a shift toward dominance by blue-green algae has been noted recently. The phytoplankton are also considerably more abundant now that the lake has become more eutrophic. Throughout the lake, blue-green algae of the genus *Aphanocapsa* can be found, occasionally in great numbers. Such "blooms" of blue-green algae (also called cyanobacteria) were observed a century ago but are more common now. Many species of this group produce lipopolysaccharides that can irritate skin and cause gastric distress in humans. As cyanobacteria have come to dominate the lake, diatoms, green algae, and dynoflagellates have become less important. Total phytoplankton biomass has reportedly increased by a factor of five since the 1960s.

The dominant species varies both seasonally and according to location. Species composition differs between inshore and open lake locations, as well as between the dry season and the rainy season. During the dry season, inshore waters are strongly dominated by cyanobacteria, but diatoms are most prevalent in the open lake. During the rainy season, cyanobacteria are still the most populous group inshore (although with lower species diversity) and outnumber diatoms in the open lake. In general, phytoplanktonic activity is greatest during the dry season. The availability of silica affects phytoplankton species diversity; during periods when concentrations of dissolved silica are low, the number of diatoms of species with high silicon requirement is also low.

Macrophytes are found in the littoral zone. Papyrus grows in the wetlands and shallow waters of the sheltered bays of Lake Victoria. Often as the papyrus population grows lakeward from the shore, it becomes detached from the bottom and forms a floating mat. These floating mats sometimes become detached in storms and can be seen around the lake. Other emergent macrophytes include cattails, reeds, and pondweed. Aquatic submerged macrophytes include coontail, hydrilla, and knotgrass. Hippo grass is an important floating macrophyte. Water hyacinth, an introduced species, has thrived in recent decades.

Invertebrates. Among Lake Victoria's zooplankton, the phyla Rotifera, Cladocera, and Copepoda encompass most of the 49 identified species. The copepods are the predominant zooplankters in Lake Victoria, making up 85 percent of the zooplankton in the Kenyan waters of the lake. Other important zooplankters include waterfleas, rotifers, and freshwater medusa. Other planktonic animals include flatworms (microturbellaria), aquatic mites (hydracarinids), and seed or mussel shrimp, a type of crustacean (ostracods). Insect larvae are also part of the zooplankton, especially larvae of the phantom midges. As adults, these nonbiting midges emerge in huge swarms and are sometimes made into "Kungu" cakes and eaten by local people.

The benthic invertebrate community has been studied primarily in the Kenyan waters of Lake Victoria. A fairly rich diversity of macroinvertebrates (nonmicroscopic animals without backbones) exists in the lake. In general, the species richness of gulfs and inshore areas is greater than that of the open waters. Oligochaete worms, molluscs, and insect larvae are well represented and widely distributed. A study of Ugandan molluscs found 65 species and subspecies of gastropods (snails) and bivalves (clams and mussels) in Lake Victoria. Lunged snails are vectors, or carriers, of a fluke that causes a serious disease (schistosomiasis) in humans and livestock.

Only one prawn occurs in Lake Victoria, the common caridina. This small shrimp-like creature occurs, or did occur, in great concentrations at times in the littoral and sublittoral zone, where it is a benthic organism. Offshore, it is a pelagic species that feeds on both detritus and plankton. Like many small pelagic animals, it exhibits a daily (diel) migration to avoid being eaten by sight-feeders (fish and other predators that feed by sight). It sinks down into darker waters during the day and then rises to the surface to feed at night. The prawn was an important prey species for some of the endemic Lake Victoria fish species, but now it seems to be preyed upon primarily by the introduced Nile perch, particularly now that the Nile perch has decimated many of the native fishes upon which it initially fed.

Reptiles and amphibians. Among the wetlands, ponds, and lakes that fringe the lake is a diverse amphibian fauna, particularly on the wetter Ugandan (western and northern) side of the lake (see Table 4.6). There are only two endemic amphibians, however, the banded toad and a frog.

Fishes. While Lake Victoria's planktonic and benthic communities harbor a great deal of biodiversity, it is the lake's fish fauna for which it is famous. Perhaps it is more accurate to say that the lake is famous for what happened to its fish fauna. The lake's native fishes are often compared to the finches of the Galapagos, made famous by Charles Darwin. Like his finches, the haplochromine cichlids of Lake Victoria are a well-known example of a species flock (see Plate VIII), which is a group of closely related species thought to have descended from a common ancestral species (that is, via adaptive radiation).

Lake Victoria's broad, shallow shape is likely partly responsible for this rapid speciation. When the lake recedes, as it does during periodic droughts, satellite lakes are cut off from the main lake. Such geographic isolation often leads to speciation. The haplochromine cichlids (cichlids belonging to the genus Haplochromis) are remarkable for their diversity. As many as 600 species of this group are thought to have evolved in Lake Victoria, in a relatively short 12,000–15,000 years at that.

Table 4.6 Amphibian Species List, Lake Nabugabo (Western Edge of Lake Victoria)

African tree frog	*Afrixalus fulvovittatus*
Egyptian toad	*Bufo regularis*
Steindachner's toad	*Bufo steindachneri*
Banded toad	*Bufo vittatus*
White-lipped hylarana	*Hyalarana albolabris*
Tree frog	*Hyalarana galamensis*
Bayon's common reed frog	*Hyperolius bayoni*
Kivu reed frog	*Hyperolius kivuensais bituberculatus*
Dimorphic reed frog	*Hyperolius cinnamomeoventris*
Long reed frog	*Hyperolius nasutus*
Common reed frog	*Hyperolius viridiflavus variabilis*
Running frog	*Kassina senegalensis*
Tree frog, common name unknown	*Leptopelis bocagei*
Puddle frog, common name unknown	*Phrynobatrachus graueri*
Snoring puddle frog	*Phrynobatrachus natalensis*
Eastern puddle frog	*Phrynobatrachus acridoides*
Disk-toed puddle frog	*Phrynobatrachus dendrobates*
Anchieta's rocket frog	*Ptychadena anchietae*
Mascarene rocket frog	*Ptychadena mascareniensis*
Rocket frog, common name unknown	*Ptychadena porississima*
Rocket frog, common name unknown	*Ptychadena oxyrrhynchus*
Chapini's common river frog	*Rana angolensis*
common name unknown	*Rana occipitalis*
African clawed frog	*Xenopus laevis victorianus*

Source: Behangana and Arusi 2004.

One aspect of the adaptive radiation of the haplochromine cichlids is their trophic, or feeding, diversity (see Table 4.7).

The adaptive radiation of Lake Victoria's cichlids into hundreds of feeding and habitat specializations gave rise to a most remarkable set of behaviors and morphological (body form) changes. Many of the haplochromines are mouth brooders. No fewer than 24 (and possibly more) are pedophages, child eaters. They eat the eggs and fry of other fishes, particularly other haplochromines. At least three, faced with the challenge of mouth-brooding mothers, evolved into "rammers," which is to say, they ram into the mothers, causing them to expel their young. Different rammers have different strategies of attack:

The Cichlids

Cichlids (pronounced *sick' lids*) are fish in the family Cichlidae. This is an enormous family of fishes whose species are distributed across Africa and the Neotropics. Some are very small, some very large. One characteristic of cichlids that is unusual among fishes is that they all give some level of parental care: guarding their eggs, in some cases even keeping the eggs in the mouth of the mother until they hatch and the fry are able to survive on their own ('mouth brooders'). In some species, just the mother is the caregiver; in others, both parents. Some cichlids are popular aquarium fish.

Table 4.7 Distribution of Lake Victoria Haplochromine Cichlids by Functional Feeding Group

Type of Feeder	Number of Species
Detritivores/phytoplanktivores	14
Phytoplanktivores	3
Epilithic grazers	5
Epiphytic grazers	14
Plant eaters	2
Molluscivore/crushers	16
Oral shellers	20
Zooplanktivores	25
Insectivores	41
Prawn eaters	13
Crab eater	1
Piscivores	114
Paedophages (prey on young of other cichlids)	24
Scale eater	1
Parasite eaters	2
Other	54
Total	349

Source: Kaufman and Ochumba 1993.

one rams from behind and hits the mother in one place; another rams from below and hits the mother in another place. A third shoots down from straight above and hits the mother on the nose. Besides the rammers, several species have evolved mouth parts that allow them to take in the mouth-brooding mother's snout, at which point the pedophage sucks out the young from the mother's mouth. A legendary haplochromine specializes in sucking the eyes out of other haplochromines, though this has not been confirmed. These behavioral adaptations are just one facet of the differentiation of these fishes. They also present a myriad of colors, shapes, and sizes (though most are relatively small). Many, if not most, will never be known to science as they became extinct before anyone knew they were there.

Besides the haplochromines, two native species of tilapiine cichlids as well as 45 noncichlid species inhabit Lake Victoria. Noncichlids include a number of cyprinids (members of the family Cyprinidae, which includes minnows and carps) and mormyrids (members of the family Mormyridae, the elephant fishes, so-called because of their elongated snouts). Almost all of Lake Victoria's mormyrids are demersal, meaning that they are bottom feeders. This specialization is likely to have worked against their survival in Lake Victoria in recent decades, as the hypolimnion there has become progressively less oxygenated. Several endemic species of catfish, including the Lake Victoria deepwater catfish, apparently met their demise in the same way, or simply by falling prey to the Nile perch.

Up until about the twentieth century, the fish fauna of Lake Victoria (or what was known of it, primarily from studying what native fishermen caught) was extraordinarily diverse and plentiful. Many traditional fishermen made a good livelihood from the lake using relatively inefficient methods of fishing. The two species of tilapias were the most important members of this native fishery in Uganda and Kenya. In the Tanzanian part of the lake, the ningu, a cyprinid, and the semutundu, a catfish, were also important. As the human population of the region increased, more efficient fishing methods, such as the use of gill nets, were introduced by the colonial administrations then in power. Fishing pressure increased. Commercial catches declined as did populations of larger fishes, one after another. From a commercial point of view, the native fishes were thought to be too bony or otherwise unsuitable, so against the advice of fisheries ecologists, the Nile perch was introduced beginning in the 1950s. The effects of this large (up to 6 ft, or about 2 m, long), aggressive predator on the native fish, in combination with overfishing, increasing pollution, and instability of lake levels, caused populations of many of the endemic fish to crash. At least 200 of the haplochromine cichlids, along with many other species, are thought to have been driven to extinction within 30 years. The many species of haplochromines once made up more than 80 percent of the fish biomass in Lake Victoria; now a single species (the Nile perch) accounts for about 80 percent of the fish biomass. If the objective was development of a commercially valuable fishery, the introduction of the Nile perch was successful. The export of Nile perch fillets from Lake Victoria is worth well over $100 million per year in a very poor region. It has not, however, proved to be a cure for the entrenched poverty that characterizes much of the region. Now the Nile perch itself is overfished and its population is declining. Another introduced species, the Nile tilapia, is flourishing and has become a commercially important species.

With the overfishing of the Nile perch, as well as some other environmental changes in the lake, there may be opportunities for some of the endemic haplochromines to rebound. The reduction in Nile perch numbers relieves some predation pressure. The choking of fishing ports by the water hyacinth has also reduced artisanal fishing pressure, and the persistent hypoxic zones that now exist most of the year in the lake may provide a refuge. Some of the native species are more tolerant of low oxygen than the Nile perch. It is a risky refuge, however; large fish kills have resulted when there is an upwelling of the hypoxic water. See the section Ecological and Environmental Changes in Lake Victoria, below, for additional information on the fishes of Lake Victoria.

Birds and mammals. Mammals and birds are often but a footnote to the dramatic story of the lake's fishes. But Lake Victoria is in the center of one of the richest areas in the world for wildlife of all kinds. Most of the megafauna that people associate with Africa—the lions, elephants, giraffes, gorillas, cheetahs, monkeys, wildebeests, and colorful birds—can be found in the region. The vast majority of these are terrestrial, but some inhabit the extensive wetlands that surround the lake.

These wetlands, certainly part of the lake's ecosystem, harbor millions of birds and mammals. The wetland complex at Sango Bay, in Uganda, is inhabited by 65 different mammals and 417 kinds of resident birds, as well as huge congregations of migratory birds. It is an important breeding ground for Squacco Herons, Grey-headed Gulls, Little Egrets, Long-tailed Cormorants, Papyrus Gonolek, Great White Pelicans, and Pink-backed Pelicans. The following bird species of concern are found at Sango Bay: Blue Swallow, Shoebill, and White-winged Black Tern. Mammals of concern at Sango Bay include the African Elephant, a subspecies of the black and white colobus monkey, the blue monkey, and the sitatunga, an amphibious antelope.

Mabamba Bay, on the northern edge of the lake in Uganda, hosts many of the same species as well as the migratory Gull-billed Terns and Whiskered Terns. Lutembe Bay's wetland complex averaged more than 1.4 million waterfowl at annual bird counts from 1999 to 2003. Most of Lake Victoria's wetlands remain unstudied, so total diversity and abundance of birds may be considerably higher.

Ecological and Environmental Changes in Lake Victoria

The food web and population dynamics of Lake Victoria, destabilized by multiple disturbances, are now in a state of flux. So many parts of the ecosystem are changing and reacting to changes in other components of the ecosystem that it is hard for scientists to keep track of, much less understand, what is happening in the lake. Continuous changes occur in land use, fishing technology, human population (expected to double within the next two decades), government policies concerning the lake, and even the climate, so the lake environment has little likelihood of reaching a stable, steady state any time soon. The actual physical environment of the lake has apparently changed. Before 1978, the lake was relatively well mixed and well oxygenated. Some time after that date, and more or less coinciding with the population explosion of the Nile perch, the lake began to stratify, a pattern which continues today. Along with stratification has come the deoxygenation of the hypolimnion, effectively eliminating up to 70 percent of the volume of the lake as fish habitat.

The species composition and food web of Lake Victoria have undergone profound changes in recent decades. These changes have been caused by nutrient loading as well as the introduction of fish species. So dramatic have been the changes that scientists view Lake Victoria with a kind of horrified fascination (more than 200 fish species have become extinct in the lake within 30 years, one of the most rapid and extensive extinctions of vertebrate species in history).

The Nile perch was introduced into the lake in the mid-1950s by officials associated with the British colonial regime in what is now Uganda. For reasons that remain unknown, the population of this fish did not increase noticeably until the period 1978–1987, when it exploded. Before the irruption of the Nile perch, 80 percent of the lake's biomass consisted of haplochromines, and afterward, 80 percent consisted of Nile perch. By the 1990s, the remaining 20 percent consisted primarily of introduced tilapia species, and whatever haplochromines were left were less than 1 percent of total biomass.

Not all the changes in the lake's food web can be attributed to predation by the Nile perch. Dramatic changes in the concentration and population dynamics of phytoplankton and zooplankton undoubtedly affect populations of organisms feeding on them. One such organism is the pelagic cyprinid known as dagaa or omena. This small fish (under 3 in, or about 8 cm) eats zooplankton and surface insects. With the near-elimination of its former predators, the haplochromines, its population exploded during the decade 1990–2000. It is now a target of some commercial fishing as well as a major food resource for cormorants and kingfishers, piscivorous birds. However, the population density of the little fish seems to be making it susceptible to a parasite, and there are concerns its population could collapse.

With haplochromines gone from the lake and their diet, the Nile perch have turned to what was once a good food source for the cichlids, the prawn. Recent surveys of the stomach contents of the Nile perch reveal that it consumes primarily prawns and juvenile Nile perch. Reportedly, some local people will not consume Nile perch because of its cannabalistic habit. But the prawn population is holding up, because the deoxygenated hypolimnion gives them a refuge from the Nile perch, which is sensitive to low oxygen levels.

Indications are that the peak population of Nile perch will not be sustained. Catches of this fish have declined since the introduction of large-mesh gill nets for the industrial offshore fishery. Nearshore, the proliferation of fine-mesh gill nets and beach seine fishing focuses on juvenile Nile perch. Nile perch population and predation has been reduced to the point at which populations of some of the surviving cichlids are beginning to grow.

Other potentially and actually invasive species have also been introduced. Several tilapia so far have not had a major impact, but the introduced Nile tilapia seems to be obliterating the two native tilapias.

The introduced water hyacinth has been a nuisance. A floating macrophyte, it was introduced to Africa early in the twentieth century. It entered Lake Victoria from the Kagera River where it was well established by the 1980s. It is now a widespread problem plant throughout the lake (as it is in many parts of the world, including the United States).

Water hyacinth grows up to three feet tall. It is a floating plant with showy lavender flowers. Its rounded leaves are leathery, its stalks are spongy, full of air for flotation. Feathery root masses hang down in the water. The plants proliferate and form tightly packed carpets that can immobilize even large boats, much less the small fishing canoes of traditional Lake Victoria fishermen. The plants block light almost completely, thus reducing phytoplankton production dramatically and affecting the entire food web. They also cover and eliminate breeding habitat for many of the lake's fishes.

At its maximum in 1998, the plant covered 50,000 ac (20,000 ha) of the lake, all in bays, gulfs, and nearshore areas. The northern rim of the lake, in Uganda and

Kenya, was affected most severely. Chemical controls (herbicides) were reportedly tried, and mechanical controls were too expensive. The use of biocontrols employing a kind of weevil seemed to have brought the water hyacinth population down, but heavy rains in 2006 brought about explosive growth again. The plant is well established in the lake and is unlikely ever to be eradicated, but decades of effort have begun to limit its further spread.

Pollution is another major problem in Lake Victoria. Concerns center primarily on eutrophication, the result of nutrient loading and buildup in the lake. Excess nutrients (nitrogen and phosphorus) can be traced to stormwater runoff from rapidly expanding urban areas, agricultural practices throughout the watershed, sewage discharge both treated and untreated, and industrial wastewater discharges. As marshes have been drained for conversion to agriculture around the lake and its tributaries, their ability to remove nutrients, particularly phosphorus, has been lost and this has accelerated the process of eutrophication. Conversion of wetlands for agricultural and urban uses not only adds sediment and nutrients but removes natural traps for both types of pollutants. An additional source of nitrate (the form of nitrogen useful to plants) in the lake is a consequence of the shift in composition of the phytoplankton. As eutrophication has proceeded, cyanobacteria have come to dominate the phytoplankton. Some are nitrogen fixers, and their abundance has added a significant amount of nitrate to the ecosystem.

Microbial pollution is another concern. Bacteria, viruses, and other microscopic organisms can cause a variety of water-borne illnesses, including cholera and dysentery. Many of the nearshore parts of the lake have high enough concentrations of such microbial pathogens to make drinking or body contact with the water dangerous. The sources of microbial contamination are not difficult to discover but may be expensive to mitigate. Of the approximately 3 million people living in urban concentrations around the lake, only about 600,000 are connected to wastewater systems that treat the wastewater before discharging it into the lake. Human waste generated by the rest goes directly into the lake or its tributaries. Another source is stormwater runoff from both urban and agricultural areas, which carries with it whatever human and animal wastes are available to be washed away when the rains come. And finally, almost none of the lake's many ships have any kind of sewage treatment, instead discharging human wastes directly into the lake.

Chemical pollution, too, is increasingly a problem. Industrial discharges are largely unregulated or regulations are unenforced. Dissolved chemicals from mining tailings find their way into the lake. Mercury from the increasing gold-mining activity in the watershed is a concern in some parts of the lake. Expired pesticides—including DDT, medical waste, and petrochemicals—all enter the lake as urbanization and industrialization increase in the watershed.

Conditions in the lake, both physical and biological, have changed dramatically since the 1960s, and rapid change continues. The lake reflects changes in the watershed, including a steady increase in human population from less than 5 million in 1950 to its current level of more than 30 million, and a doubling of agricultural output. Whether and how the ecosystem of Lake Victoria will stabilize is a question scientists are struggling to answer. The answers are critical to the well-being of millions of people in the region who depend on the lake for their livelihoods and basic needs.

With respect to water quality, since the 1960s Lake Victoria has become progressively more eutrophic. One typical consequence of increasing eutrophication is diminished light penetration. In the 1960s, the Secchi depth in the lake was measured at 23 ft (7 m); by the 1990s, it was down to about 6 ft (2 m). In the late 1990s another study found Secchi depth in one of the large bays in the Kenyan part of the lake to be 1.6–3.3 ft (0.5–1 m) and in the open lake to be 3–8 ft (0.9–2.4 m). Secchi readings have always been higher and water clearer in the open lake than in the bays, gulfs, and littoral zone generally. They are lower during algal blooms, so there is short-term variation amid long-term trends. While there is no doubt that the lake is becoming more eutrophic, as recently as 1991 a Secchi reading of about 200 ft (60 m) was recorded in the open lake.

Chlorophyll *a* levels, which are used to measure the concentration of photosynthecizing algae, increased fivefold from the 1960s to the early 1990s. Primary production, another measure of algal activity, increased fourfold. Prior to the 1960s, the low oxygen levels were rare except below about 200 ft (60 m) of depth. Low oxygen is now common even in shallow waters, a typical consequence of increased algal populations. The algae die and sink to the depths, where they decompose. The decomposition process uses oxygen, and with stratification, the waters are not reoxygenated. Hypoxic (low-oxygen) and anoxic (no oxygen) water is unsuitable habitat for fish and most other aquatic organisms, which now have approximately 70 percent less habitat than they had 50 years ago. Occasional upwellings of hypoxic water cause massive fish kills on Lake Victoria.

The change in water chemistry driving the eutrophication is a dramatic increase in loadings of phosphorus and nitrogen to the lake. Phosphorus inputs are now three times what they were in the 1960s, and nitrogen loading has increased also. Concentrations of nitrate and phosphate in 1988 were similar to their concentrations in the 1960s, the excess having been taken up by algae production. Nitrogen is now considered to be the limiting nutrient, so further increases in nitrogen will lead to further increases in algae concentration. There is now evidence that algae concentrations are so high that another factor may begin to limit further population growth of algae: light availability. The higher the algae concentration, the smaller the euphotic zone; the smaller the euphotic zone, the fewer algae can live there.

The levels of nutrients dissolved in the lake's waters, and the populations of algae, are dynamic. Lake Victoria is monomictic, that is, its water mixes vertically once a year, at the end of July when the lake reaches temperature uniformity throughout (that is, it becomes isothermal). When this mixing occurs, nutrients from the hypolimnion are brought to the euphotic zone again, and algal blooms can occur. Nitrogen levels are also affected by a change in the composition of the algal community, which is now dominated by species of blue-green algae that can fix nitrogen (that is, transform atmospheric nitrogen into a form usable by plants).

Silica concentration in the lake was reduced by 90 percent between the 1960s and 1988, apparently because of increased diatom production. Silica is an important nutrient for diatoms, which are encased in glass shells.

Lake Victoria's Prospects

The watershed is home to at least 30 million people, of whom two-thirds rely on subsistence agriculture or grazing animals for their livelihoods. The region has some of the highest rates of population growth in the world, averaging over 3 percent per year. Half of the population is under the age of 15, which means that considerable population growth lies ahead. The region is also one of the world's poorest as measured by per capita income, and food security is an issue for many (in a region that exports tens of thousands of tons of high-quality protein in the form of fish fillets each year).

Land use in the watershed is primarily agricultural, mostly small-scale subsistence production. Expansion of croplands and overgrazing are blamed for much of the water pollution (nutrients, sediment, pesticides) that currently affects the lake. In 2003, approximately one-third of the watershed was under cultivation, a proportion that has undoubtedly increased since. Agricultural conversion of wetlands around the lake and especially in the tributary watersheds is cited as a cause of water-quality degradation.

The fishes of Lake Victoria are receiving a great deal of attention from scientists and officials of government, nongovernmental organizations, and international organizations, including the World Bank and the United Nations. Studies are under way and efforts are being made to improve environmental conditions in the lake and its watershed. However, the future is uncertain for Lake Victoria and its fabled fishes. Conditions in the basin are changing rapidly. Wildlife habitat, including wetlands, is being converted to farmland, as communities work to feed a rapidly growing population of people and cattle. Thousands of tons of fish fillets are exported onto world markets, while protein deficiency among local people is increasing. At the same time, however, ecotourism is booming, giving governments an economic rationale for nature preservation. But preservation, if it comes, will come too late for many of Lake Victoria's fishes.

In all, Lake Victoria's ecosystem is a system out of balance: "A community of more than 400 fish species collapsed to just three co-dominants: the Nile perch, the Nile tilapia, and a single indigenous species, the omena" (Kaufman 1992). Diversity of species and complexity in food webs usually promotes stability and resilience; simplicity in food webs and reliance on few species make an ecosystem unbalanced and unstable.

A Temperate, Heavily Modified Lake: Lake Ontario

The North American Great Lakes

The Great Lakes make up North America's greatest lake system. Three of the five major lakes in the system are among the 10 largest freshwater lakes (by surface

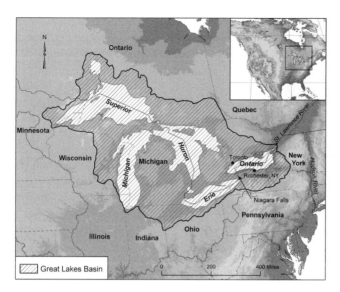

Figure 4.6 The North American Great Lakes and their watershed. *(Map by Bernd Kuennecke.)*

area) in the world. Some consider Lake Michigan and Lake Huron to be a single lake, since the two are connected hydrologically, exchange water freely, and therefore have the same water level. Each lake individually is large enough to be included among the 10 largest freshwater lakes. The Great Lakes hold approximately a fifth of the world's fresh water and together their surface area (94,270 mi^2 or 244,160 km^2) is larger than that of the United Kingdom (England, Wales, Scotland, and Northern Ireland).

The land area of the Great Lakes watershed (see Figure 4.6), at a little over 200,000 mi^2 (521,000 km^2), is more than twice as large as the surface area of the lakes themselves. A little more than half of this land area is in the United States, with the balance in Canada. Eight states and one province (Ontario) share more than 10,500 mi (17,000 km) of shoreline.

The Great Lakes are located in an area of mid-latitude humid continental climate (Koeppen classifications Dfa, Dfb). Characteristics of such climates include relatively low annual precipitation, four distinct seasons, and great seasonal variability with hot summers and cold winters. Superimposed on this climate template, however, is the moderating effect of the sheer mass of the lakes themselves. Because the lakes' water does not warm or cool quickly, seasonal extremes of temperature are lessened nearby. The large water surface area results in greater humidity overall. It is this humidifying effect that results in the "snow belt" on the southern and eastern shores, which are downwind of the water surfaces with respect to prevailing winter winds. The temperature in all seasons is cooler in the northern basin and warmer in the southern. Only Lake Erie, more southern in

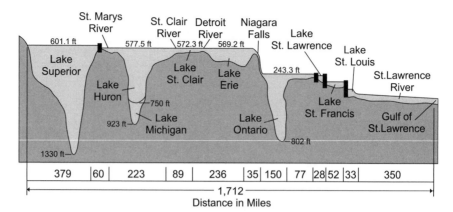

Figure 4.7 Great Lakes elevation profile. *(Illustration by Jeff Dixon. Adapted from a U.S. Army Corps of Engineers illustration in University of Wisconsin Sea Grant Institute, n.d.)*

location but much shallower than the other lakes, freezes over completely, but it does not do so every year. The great mass of water in the other lakes, relative to surface area, prevents them from freezing over completely.

The Great Lakes are of glacial origin. The Laurentide Ice Sheet, in a relatively recent glaciation (10,000–15,000 years ago), gouged out the lakes' basins. Ice sheets more than 1 mi (1.6 km) thick ploughed up soil and relatively soft rock, damming a preexisting river system in the process. When the glaciers melted, their meltwater filled the basins they had created.

The Great Lakes are linked together by water, which flows from Lake Superior and Lake Michigan, into Lake Huron. Lake Huron's discharge flows into Lake Erie; Lake Erie's discharge flows, via the Niagara River, into Lake Ontario. The entire system's discharge flows to the sea via the Saint Lawrence River. Figure 4.7 shows the Great Lakes' relation to one another in profile, and Table 4.8 presents morphological data for the Great Lakes.

Physical Characteristics of Lake Ontario

Lake Ontario serves as an example of a temperate climate lake system. Like many of the world's large lakes, Lake Ontario supports a wide variety of uses and provides many benefits to society: fisheries; recreation; transportation; domestic, agricultural, and industrial water; and waste assimilation. In many ways, however, the lake's ability to continue to provide some of those benefits is in question. As is the case with many of the world's lakes, Lake Ontario is ecologically in trouble. Pollution and introduced exotic species threaten the lake's ecosystem. Climate change, signs of which are already evident throughout the Great Lakes, will further alter the system in unpredictable ways.

Lake Ontario is the smallest of the Great Lakes by every surface measure (length, width, area, shoreline), but it is relatively deep and therefore has more than

Table 4.8 Great Lakes Morphometry

Lake	Mean Elevation [ft (m) a.s.l.]	Area [mi² (km²)]	Volume [mi³ (km³)]	Length [mi (km)]	Width [mi (km)]	Shoreline [mi (km)]	Mean Depth [ft (m)]	Maximum Depth [ft (m)]
Lake Superior	600 (183)	31,820 (82,414)	2,945 (12,233)	350 (563)	160 (257)	2,979 (4,795)	489 (149)	1,329 (405)
Lake Michigan	579 (176)	22,400 (58,016)	1,179 (4,913)	307 (494)	118 (190)	1,636 (2,633)	279 (85)	922 (281)
Lake Huron	579 (176)	23,010 (59,596)	844 (3,516)	206 (332)	183 (295)	3,826 (6,157)	194 (59)	751 (229)
Lake Erie	570 (174)	9,940 (25,745)	117 (488)	241 (388)	92 (57)	871 (1,402)	62 (19)	210 (64)
Lake Ontario	245 (75)	7,540 (19,529)	391 (1,631)	193 (311)	53 (85)	712 (1,146)	282 (86)	801 (244)

three times the volume of Lake Erie (see Table 4.8). Lake Ontario is the lowest of the Great Lakes in terms of surface elevation, and the most downstream of them. It is the only one of the Great Lakes with a naturally occurring anadromous fish species (the Atlantic salmon), as Niagara Falls from time immemorial has presented an insurmountable obstacle to upstream migration.

The climate of the Lake Ontario basin is characterized as Dfa in the Koeppen climate classification system: humid with a severe winter, no dry season, and a hot summer. In winter, the basin is dominated by cold arctic airmasses, while in summer, warm humid airmasses from the Gulf of Mexico predominate. Mean January temperature around the lake is from $0°–27.5°$ F ($-18°$ to $-2.5°$ C); mean temperature in July is $65°–70°$ F ($18°–21°$ C). Mean annual precipitation varies from around 31.5 in (800 mm) in the western part of the basin to 40 in (1,000 mm) in the east, parts of which can be characterized as a snowbelt. The lake's moderating influence on temperatures has encouraged the development of orchards in the Lake Ontario region of New York State.

The lake is relatively narrow and deep, with two major basins separated by a low 131 ft (40 m) rise in the lake bottom extending northward from around Rochester, New York. The eastern basin, known as the Rochester Basin, contains the lake's greatest depths. The bottom drops away sharply on the south (New York) side of the lake and more gradually on the north side.

Lake Ontario's watershed (not counting the watersheds of the upstream Great Lakes) is home to 7.6 million people, of whom 5.4 live on the Canadian side. The largest concentration of people is in the so-called Golden Horseshoe, extending from the Niagara Falls area around the west end of the lake to the Toronto metropolitan area. There are a few smaller urban concentrations on the U.S. side, notably Rochester and Oswego, New York. Land use on both the Canadian and U.S. sides of the lake is dominated by agriculture and forestry (39 and 49 percent of the basin, respectively). The U.S. side has proportionately more forest and less agriculture than Canada. Urban and other uses make up a relatively small percentage of the basin's land use, but residential and commercial land uses are more evident in the part of the basin closest to the lake.

Water movement and residence time. Hydrologically, the lake's water volume is large in relation to the magnitude of its inflows and outflows, but much less so than some of the other Great Lakes. It would take seven years for the current inflows into the lake to fill it (which is to say that it has an average residence time of seven years); Lake Superior, by contrast, would take almost 200 years to fill. Eighty percent of the inflow to Lake Ontario comes from the Niagara River, which delivers water from the upstream chain of Great Lakes at an average rate of 200,000 ft^3/sec (5,663 m^3/sec). The remainder comes from tributaries in the lake's basin (14 percent) and precipitation falling directly on the lake (7 percent). The lake's water exits via the Saint Lawrence River (93 percent) and by evaporation (7 percent).

The construction in the 1950s of the Saint Lawrence Seaway, an extensive system of navigational locks, dams, and canals, has had numerous impacts upon Lake

Ontario. Indirectly, the admitting of oceangoing vessels into the Great Lakes system has resulted in numerous nonindigenous species introductions that are profoundly affecting the native ecosystem. A direct effect is the regulation of lake level, which in Lake Ontario used to vary over a range of about 6.5 ft (2 m), but now varies less than half of that.

Temperature and mixing regime. Lake Ontario is dimictic, that is, the waters mix vertically twice during the course of a year. Vertical stratification begins to develop mid-June to July, with a marked thermocline forming in the offshore areas at depths of 30–50 ft (about 10–15 m), although this varies from one part of the lake to another. The stratification of the lake into a warm, nutrient-deprived but oxygen-rich epilimnion and a cool, nutrient-rich but oxygen-deprived hypolimnion begins to break down in September. Surface temperatures gradually cool until the epilimnial temperature approaches that of the hypolimnion, which stays at around 39° F (4° C) throughout the year.

As winter sets in, surface temperatures cool to below 39° F (4° C); water is most dense at 39° F (4° C), so the cooler water "floats" on top of the denser water. Partial icing of the lake's surface reduces wind-driven mixing, and stratification occurs again, though not as complete as in the summer.

Even in the summer, some mixing does occur. A flow of water is present where the Niagara River enters the lake. More important, the prevailing westerly winds push the surface waters to the east and (because of the Coriolis effect) to the south. As the lake water "piles up" on the southern and eastern shores, the surface water sinks. On the northern and western shore, there is a corresponding upwelling of hypolimnial water. In addition, depending on season and other conditions, lake currents exist that either proceed in a counterclockwise loop around the lake's perimeter, or form two gyres—a counterclockwise one in the eastern basin (the Rochester basin) and a clockwise one in the western basin.

Another phenomenon associated with mixing and the transition to summer stratification is the formation of a horizontal thermocline. In spring, shallower water around the lake's edges warms first, and rises by convection. As it spreads over the surface, it flows away from the shore. A line or "thermal bar" appears where this spreading warm surface water meets the still-cold surface water offshore; the mixing of these two water masses produces a zone of maximum density, at which surface water from both directions sinks downward. This zone of sinking moves like a concentric ring slowly toward the lake's center, eating into the cold offshore surface water and replacing it with warmer surface water. Eventually the warm surface water overspreads the entire lake, and at that point the lake begins to stratify vertically. In the meantime, as the thermal bar moves outward away from the shore, nutrient-rich deeper water is pulled up to replace the surface water, and this upwelling and warming are associated with explosive increases in phytoplankton activity.

Summer stratification strongly affects the biological conditions in the lake. The epilimnion becomes sealed off from the hypolimnion. Nutrients in the epilimnion are used by phytoplankters. As these and other organisms near the surface die and sink, nutrients are removed from the epilimnion, which becomes progressively more nutrient deprived. Meanwhile the accumulation of dead organic material in the hypolimnion increases decomposition, and this process consumes oxygen. As noted above, however, the stratification is incomplete and some mixing does occur, resulting in periodic reintroduction of nutrients to the epilimnion.

Some mixing is also caused by the seiche movement of water (sometimes called "slosh" on the North American Great Lakes). Lake Ontario has a seiche period of about 11 minutes; normally the amplitude of the seiche is barely detectable, about 0.8 in (2 cm). However, atmospheric conditions such as an advancing high-pressure ridge can greatly increase the magnitude of the seiche.

Ice regime. The lake rarely freezes over, but inshore areas freeze, particularly on the south shore and the Kingston basin (the northeastern corner, where the lake tapers to the Saint Lawrence River outlet). Ice coverage tends to peak in February. The average annual ice coverage maximum is in the range of 20 percent with considerable variability; in 1979, ice coverage peaked at 80 percent of the lake's surface.

Water chemistry. Major ions in Lake Ontario's water include the cations calcium, magnesium, sodium, and potassium and the anions carbonate/bicarbonate, sulfate, and chloride. The epilimnion is occasionally supersaturated with calcium carbonate, leading to precipitation of calcium carbonate crystals that sink into the hypolimnion. Concentrations of all of the major ions increased during the period 1850–1970 as a result of human activity in the basin and in upstream basins, particularly that of Lake Erie.

Trophic status. The "natural state" of Lake Ontario is oligotrophic—low levels of plant life, low nutrient concentrations. During the period 1950–1975, the lake became mesotrophic overall and even eutrophic in nearshore areas because of increased phosphorus and nitrogen loadings from both point and nonpoint sources. Both available nitrogen and total phosphorus averaged more than 20 µg/L during the period 1968–1975, for example. As a result of a concerted effort by the governments of Canada and the United States, levels of these nutrients were in decline by the 1980s, and the lake was trending toward an oligotrophic state again. As of 2006, it was not clear whether the efforts to reduce phosphorus had simply been too effective, or whether the trend toward a too-oligotrophic state was brought about by two invasive exotic species, the zebra mussel and the quagga mussel, both of which have multiplied prolifically and spread throughout the Great Lakes.

Biota of Lake Ontario

Macrophytes. Except in some sheltered bays and "ponds" around the shoreline, bottom-rooted vascular plants make relatively little contribution to overall production in the Lake Ontario ecosystem. However, areas where higher aquatic plants grow are important spawning or nursery habitat for a number of fishes, including some that only spawn among aquatic plants, such as the northern pike, muskellunge, and gar.

In sheltered bays protected from the effects of wave energy, the dominant macrophytes are tape grass, pondweeds, naiad, horned pondweed, water stargrass, and coontail.

Elevated nutrient levels in some of these protected bays have led to abundant growth of filamentous algae and muskgrass, also an algal species. Other aquatic macrophytes have suffered the effects of profuse growth of these algae in the nearshore area. Even now, as the lake overall has become much more oligotrophic, the problem persists because of the higher average concentrations of both nitrogen and phosphorus in the nearshore zone. When filamentous algae dies, it washes up on beaches, creating an aesthetic problem. Even on rocky shores exposed to wind and wave action, where other macrophytes are rare, it can be found.

Wetlands are found in some of the sheltered shoreline areas around Lake Ontario. Along the southern (New York) shore of the lake, the lake intercepts an undulating topography, creating embayments. Low barrier beaches across their mouths have created shallow lagoons. Barrier beach lagoons and associated wetlands also are found along the eastern end of the lake. Emergent vegetation in the lowest elevations of these wetlands include the common hornwort or coontail and common waterweed, as well as duckweeds. The floating-leaved yellow lily and fragrant water lily are also common. These plants prefer nutrient-rich, low-energy conditions found in these protected lagoons. In cooler, spring-fed wetlands, high densities of ivy-leaf duckweed may be found, along with pondweed.

Upslope from the emergent zone, wet meadow can typically be found, dominated by cattails, bluejoint reedgrass, and marsh fern. The dominance of cattails today reflects the stabilization of lake level associated with the Saint Lawrence Seaway. Farther upslope, a shrubby zone marks the transition to dry land.

Phytoplankton. Phytoplankton populations respond robustly to changes in nutrient levels, particularly of nitrogen, phosphorus, and (to a lesser degree) silicon. Their populations also respond to water temperature and light availability; thus there is a strong seasonality to concentrations of phytoplankton. Although there is considerable variation from year to year, there is generally a spring algal bloom and a late summer bloom. Spring, summer, and fall are dominated by different species. The diatoms of the genera *Cymbella* and *Stephanodiscus* are important in spring, while the late summer bloom is dominated by green algae (division Chlorophyta), particularly *Phacotus lenticularis*, *Oocystis* spp., *Staurastrum paradoxum*, and *Ulothrix subconstricta*. Seasonality also affects the distribution of different species across the lake, with the thermal bar separating phytoplankton communities of differing compositions.

The species mix has changed over time in response to changing environmental conditions in the lake. During the nutrient-rich 1970s, it was found that, on an annual basis, the total phytoplankton biomass (measured as carbon) was composed of 57.7 percent diatoms, 17.0 percent green algae, 13.1 percent dinoflagellates, and 5.3 percent flagellates. The remaining 7.2 percent was made up of cyanobacteria (blue-green algae), cryptomonads, chrysophytes, and euglenoids.

In 2003, total phytoplankton biomass was low compared with previous years, perhaps a tenth of the amount measured in 1990. But the cyanobacteria, or blue-green algae, were found to make up nearly half of the summer biomass of phytoplankton. This has alarmed fisheries biologists and lake managers because the blue-green algae are a poor-quality food source for zooplankton, which in turn are consumed by many fish species. The generally low levels of phytoplankton have improved water clarity, which in turn has led to a dramatic increase in filamentous algae, which attaches to the benthos, growing in the nutrient-rich nearshore areas.

Differences in spatial distribution of phytoplankton species, both vertical and horizontal, are not unexpected considering the variations in water temperature and nutrient concentrations both around the lake and throughout the course of the year. In the vertical profile, phytoplankton densities peak between 16 and 33 ft (5 and 10 m) depth in offshore areas of the lake. (Much of the basic knowledge concerning species distribution, species mix, and phytoplankton concentration is from studies conducted in the 1970s and 1980s. However, the rate of change in Lake Ontario's ecosystem is such that many of the findings of the 1970s studies may no longer be relevant.)

Zooplankton. As is the case with most lakes, the zooplankton of Lake Ontario is dominated by three groups: protozoans, rotifers, and crustaceans. While the protozoans and rotifers occur seasonally in great numbers, their biomass and therefore their importance in the food web is small compared with the crustacean zooplankters.

In a 1998 survey, zooplankton biomass in Lake Ontario was dominated in the spring by cyclopoid copepods, a group of microscopic crustaceans of subclass Copepoda. Calanoid copepods were also represented. By summer, the zooplankton was more diverse. The three groups making up the largest proportion of biomass were nauplii (a collective term for the larvae of many crustaceans) and water fleas (Daphnia and Bosmina, both crustaceans of order Cladocera). Also represented were the calanoid and cyclopoid copepods, and rotifers.

The relatively large biomass of zooplankton indicates a plentiful supply of food, as would be usual in a relatively eutrophic lake. As the lake trends toward oligotrophic status, shifts in the abundance of crustacean species and overall abundance among the zooplankton are expected. In 2003, this expectation was partially confirmed as both density and biomass of zooplankton in the epilimnion had decreased substantially in all three seasons sampled. This probably reflects a bottom-up effect of reduced phytoplankton abundance caused by the increasing numbers of the efficient filter feeders—the zebra mussel and especially the quagga mussel. As these invaders continue to spread across the lake bottom, further changes in the zooplankton can be expected.

A top-down effect of predation on zooplankton abundance and species composition is being seen as a result of the introduction of two nonindigenous crustaceans, the spiny water flea and a calanoid copepod, *Eurytemora affinis*. The spiny water flea is an unusually large cladoceran (up to 0.4 in or 1 cm long) and a ferocious predator. Two of its favorite prey are important zooplankters in Lake Ontario: the cladocerans *Bosmina longirostris* and *Ceriodaphnia lacustris*. Biologists are concerned about the impact of the spiny water flea and another, similar nonindigenous invader, the fishhook waterflea, because unlike the zooplankters they are eating, these exotics are not particularly attractive prey for young fish because of their long spiny tails.

Most zooplankters in Lake Ontario reach peak abundances in late summer. Temperature seems to be a major control on population: hardly any can be found during the cooler months and higher densities are found in the warm months in parts of the lake where epilimnial temperatures are higher (generally the eastern end of the lake).

As the waters of the lake stratify, the zooplankton also stratify. Most of the crustacean zooplankton are epilimnial. A few of the larger ones prefer cooler waters and therefore are found in the hypolimnion during stratification: two calanoid copepod species and a mysid shrimp. The latter two species move diurnally up through the water column to feed in the epilimnion at night and return to the hypolimnion during the day. This is an adaptation to avoid predation by fish. The mysid shrimp is an important link in the food web between the detritus-based benthic community and the fishes and other organisms of the pelagic zone.

Benthos. The macroinvertebrate community dwelling on the lake bottom is composed primarily of two groups, amphipods, which are a kind of crustacean, and oligochaetes, which are segmented worms similar to earthworms. In the 1970s, the oligochaetes accounted for 56 percent and the amphipods accounted for 36 percent of organisms in the samples. More recent surveys have found that the amphipod genus, *Diporeia*, make up about 50 percent of the benthos in Lake Ontario, whereas oligochaetes worms make up only 30 percent. Other species present include various freshwater clams and midge larvae, sphaeriid clams, and snails. Recently the invasive, nonindigenous zebra mussel and the quagga mussel have come to dominate the benthos.

The benthic community in the littoral zone, in waters less than 35 ft (about 10 m) deep, is still dominated by oligochaetes and amphipods, but of different species. Amphipods of genus *Gammarus* predominate, and instead of chironomid larvae being the only insect representatives, there are a number of other insects as well, including mayflies, caddisflies, stoneflies, dobsonflies and alderflies, dragonflies, and damselflies.

By the early 1990s, the zebra mussel was by far the most abundant benthic organism in the littoral zone, but surprisingly, its presence seemed beneficial to the other benthic macroinvertebrates, which all became more abundant than in the

Zebra and Quagga Mussels

The zebra mussel first appeared in the Great Lakes in 1988, transported from the Caspian Sea region of Eurasia in an oceangoing vessel in its larval stage, and discharged along with ballast water. Only a year later, quagga mussels followed. The mussels spread quickly through the Great Lakes and their populations soon reached astronomical numbers. Zebra mussels have spread into the Mississippi and Ohio River systems as well as the Hudson and are considered a major threat to North America's unique and diverse native freshwater mussel fauna.

In the Great Lakes, the impact of the zebra mussel has been profound; that of the quagga mussel is just beginning to be recognized. Filter feeders, zebra mussels consume huge quantities of phyto- and zooplankton and have had a tremendous impact on the Great Lakes' food web. They foul shipping gear as well as navigational and water intake structures. They also attach to crayfish, native mussels, turtles, and other zebra mussels. Their ecological impact, via their consumption of plankton, is of great concern. Although several diving ducks and fishes eat them, none does so enough to even begin to control their numbers.

In Lake Ontario, the zebra mussel seems to have been superseded by the quagga. Zebra mussels prefer relatively shallow water (6–26 ft or 2–8 m), which in Lake Ontario is a relatively small percentage of the lake bottom. They also tend to limit themselves to hard surfaces, but the lake bottom is mostly composed of soft sediments. The quagga, however, has recently been found carpeting the bottom down to depths greater than 300 ft (about 90 m), and it is undeterred by soft substrates. As its population explodes in Lake Ontario, that of the zebra mussel is declining.

more eutrophic conditions of the 1970s. Whether this improvement in conditions will be seen as the quagga mussel replaces the zebra mussel is doubtful.

In Lake Ontario, the most worrisome impact of the two imported mussel species, particularly of the quagga, is on the food web through its effects on the benthic amphipod *Diporeia*. *Diporeia* is an important prey species for a number of fish, including the native lake whitefish. The expansion of the quagga into ever-deeper waters has been accompanied by a deepening decline in *Diporeia* densities, apparently as the result of competition for food (phytoplankton).

Fishes. The assemblage of fishes of Lake Ontario is the product of both millennia of natural processes and 200 years of human impacts, most important, fishing, pollution, and introduction of nonindigenous species. Because of the geological youth of the Great Lakes, there have been relatively few endemics, and several of them are already extinct. For the most part, the original fish fauna of the Great Lakes in general, and Lake Ontario in particular, consisted of species from the Mississippi River system, with a smaller number from the Susquehanna and Hudson River systems. Nothing comparable to the endemic species flocks of the African Great Lakes occurs in the young Great Lakes of North America. Still, at the time of European settlement of the watershed, there were 116 fish species in Lake Ontario.

Distinctive assemblages of fishes occupy the several habitats of Lake Ontario. Coastal nearshore areas (less than 50 ft/15 m deep) have relatively few species, and most of these are also found offshore. Alewives and indeed most of the lake's fish use the nearshore area either for spawning or as a nursery area, or both. The larger bays, including the Bay of Quinte and the Outlet Basin in the eastern lake, have greater diversity. In these bays, the top predators are longnose gar, bowfin, northern pike, smallmouth and largemouth bass, and walleye. A variety of other species, including gizzard shad, white sucker, brown bullhead, American eel, trout-perch, white perch, yellow perch, and freshwater drum, along with a number of minnows and sunfishes are found in these bays.

A second major habitat in Lake Ontario is the benthic offshore habitat: habitat on or near the bottom well offshore, in relatively deep to very deep water. Fishes living close to the bottom are called demersal fish. Relative abundance of fishes in this zone, as in other parts of Lake Ontario, is extremely unstable. Under the best of conditions, fish populations are dynamic, and in a lake in which fishes are subject to as many environmental changes as Lake Ontario, the relative numbers are difficult to predict from one year to the next. The piscivorous lake trout, when it is present in significant numbers, is the top predator in this habitat zone. Lake whitefish and the slimy sculpin feed on benthic macroinvertebrates, and the slimy sculpin is an important prey fish for juvenile lake trout. Less abundant benthic fishes include deepwater sculpin, round whitefish, and burbot. Food webs of the benthic offshore zone and the pelagic offshore zone are linked by the vertical migrations (in some cases daily) of mysids (the "krill" of the Great Lakes), alewives, and rainbow smelt.

The offshore pelagic zone is home to large predatory fish, many introduced and maintained for the benefit of anglers through stocking. In this category fall Chinook salmon, coho salmon, rainbow trout, brown trout, and adult lake trout. These fishes consume smaller prey that once included lake herring, deepwater cisco, and whitefish as well as several sculpins. Now their prey is limited mostly to the introduced alewife and the possibly introduced rainbow smelt. The prey fish, or "forage" fish, mostly feed on zooplankton, primarily copepods and cladocerans. Other inhabitants of the offshore pelagic zone include the threespine stickleback, emerald shiner, and gizzard shad. A complete list of Lake Ontario fish species is given in Table 4.9.

Birds. Thousands of resident birds inhabit Lake Ontario and its shoreline habitats. Large colonies of waterbirds once feasted on the abundant fish in the lake. Their numbers were much reduced by pollution and habitat loss in the middle of the twentieth century. Important waterbird species include the Common Loon, Common and Caspian Terns, Horned Grebe, Mallard, Long-tailed Duck, Double-crested Cormorant, Red-breasted Merganser, Bonaparte's Gull, Ring-billed Gull, Herring Gull, and Great Black-backed Gull.

Lake Ontario is a stop for many of the migratory birds using the Atlantic flyway. Tens of thousands of ducks, geese, passerines, and neotropical migrants use the lake and particularly its wetlands as a stopover.

Environmental Problems of Lake Ontario

Lake Ontario's ecological community is suffering the effects of diverse environmental insults, and although improvements have been made in some areas, a number of areas of concern remain. These focus primarily on the effects of habitat loss

Table 4.9 Lake Ontario Fish Species

Alewife	*Alosa pseudoharengus*
American brook lamprey	*Lampetra appendix*
American eel	*Anguilla rostrata*
American shad (extirpated)	*Alosa sapidissima*
Atlantic salmon	*Salmo salar Linnaeus*
Banded killifish	*Fundulus disphanus*
Bigeye chub	*Notropis amblops*
Bigmouth shiner	*Notropis dorsalis*
Black bullhead	*Ictalurus melas*
Black crappie	*Pomoxis nigromaculatus*
Blackchin shiner	*Notropis heterodon*
Blacknose dace	*Rhinichthys atratulus*
Blacknose shiner	*Notropis heterolepis*
Blackside darter	*Percina maculata*
Bloater (extirpated)	*Coregonus hoyi*
Blue Pike (extinct?)	*Stizostedion vitreum glaucum*
Bluegill	*Lepomis macrochirus Rafinesque*
Bluespotted sunfish (introduced)	*Enneacanthus gloriosus*
Bluntnose minnow	*Pimephales notatus*
Bowfin	*Amia calva Linnaeus*
Bridle shiner	*Notropis bifrenatus*
Brindled madtom	*Noturus miurus*
Brook silverside	*Labidesthes sicculus*
Brook stickleback	*Culaea inconstans*
Brook trout	*Salvelinus fontinalis*
Brown bullhead	*Ictalurus nebulosus*
Brown trout (introduced)	*Salmo trutta*
Burbot	*Lota lota*
Common carp (introduced)	*Cyprinus carpio*
Central mudminnow	*Umbra limi*
Central stoneroller	*Campostoma anomalum*
Chain pickerel	*Esox niger*
Channel catfish	*Ictalurus punctatus*
Channel darter	*Percina copelandi*
Chinook salmon	*Oncorhynchus tshawytscha*
Cisco or lake herring	*Coregonus artedii*
Coho salmon (introduced)	*Oncorhynchus kisutch*
Common shiner	*Notropis cornutus*
Creek chub	*Semotilus atromaculatus*
Creek chubsucker	*Erimyzon oblongus*
Cutlips minnow	*Exoglossum maxilingua*
Deepwater sculpin	*Myoxocephalus thompsoni*
Eastern sand darter	*Ammocrypta pellucida*
Eastern silvery minnow	*Hybognathus regius*
Emerald shiner	*Notropis atherinoides*

Fallfish	*Semotilus corporalis*
Fantail darter	*Etheostoma flabellare*
Fathead minnow	*Pimephales promelas*
Finescale dace	*Phoxinus neogaeus*
Fourhorn sculpin	*Myoxocephalus quadricornis*
Freshwater drum	*Aplodinotus grunniens*
Gizzard shad	*Dorosoma cepedianum*
Golden redhorse	*Moxostoma erythrurum*
Golden shiner	*Notemigonus crysoleucas*
Goldfish	*Carassius auratus*
Grass pickerel	*Esox americanus*
Greater redhorse	*Moxostoma valenciennesi*
Green sunfish	*Lepomis cyanellus*
Greenside darter	*Etheostoma blennioides*
Hornyhead chub	*Nocomis biguttatus*
Iowa darter	*Etheostoma exile*
Johnny darter	*Etheostoma nigrum*
Kiyi (extirpated)	*Coregonus kiyi*
Lake chub	*Couesius plumbeus*
Lake chubsucker	*Erimyzon sucetta*
Lake herring	*Coregonus artedi*
Lake sturgeon	*Acipenser fulvescens*
Lake trout	*Salvelinus namaycush*
Lake whitefish	*Coregonus clupeaformis*
Largemouth bass	*Micropterus salmoides*
Least darter	*Etheostoma microperca*
Logperch	*Percina caprodes*
Longear sunfish	*Lepomis megalotis*
Longnose dace	*Rhinichthys cararactae*
Longnose gar	*Lepisosteus osseus*
Longnose sucker	*Catostomus catostomus*
Margined madtom	*Noturus insignis*
Mimic shiner	*Notropis volucellus*
Mooneye	*Hiodon tergisus*
Mottled sculpin	*Cottus bairdi*
Muskellunge	*Esox masquinongy*
Ninespine stickleback	*Pungitius pungitius*
Northern hog sucker	*Hypentelium nigricans*
Northern pike	*Esox lucius*
Northern redbelly dace	*Phoxinus eos*
Pearl dace	*Margariscus margarita*
Pink salmon (introduced)	*Onchorhynchus gorbuscha*
Pirate perch	*Aphredoderus sayanus*
Pugnose shiner	*Notropis anogenus*
Pumpkinseed	*Lepomis gibbosus*

(Continued)

Table 4.9 (*Continued*)

Quillback	*Carpiodes cyprinus*
Rainbow darter	*Etheostoma caeruleum*
Rainbow smelt	*Osmerus mordax*
Rainbow trout (introduced)	*Salmo gairdneri*
Redfin pickerel	*Esox americanus*
Redfin shiner	*Notropis umbratilis*
Redside dace	*Clinostomus elongatus*
River chub	*Nocomis micropogon*
Rock bass	*Ambloplites rupestris*
Rosyface shiner	*Notropis rubellus*
Round whitefish	*Prospium cylindraceum*
Rudd (introduced)	*Scardinius erythrophthalmus*
Sand shiner	*Notropis stramineus*
Satinfin shiner	*Notropis analostanus*
Sauger	*Stizostedion canadense*
Sea lamprey (introduced?)	*Petromyzon marinus*
Shorthead redhorse	*Moxostoma macrolepidotum*
Shortnose cisco (extinct?)	*Coregonus reighardi*
Silver chub	*Hybopsis storeriana*
Silver lamprey	*Ichthyomyzon unicuspis*
Silver redhorse	*Moxostoma anisurum*
Silvery minnow	*Hybognathus nuchalis*
Slimy sculpin	*Cottus cognatus*
Smallmouth bass	*Micropterus dolomieui*
Splake	*Salvelinus fontinalis*
Spoonhead sculpin (extirpated)	*Cottus ricei*
Spotfin shiner	*Notropis spilopterus*
Spottail shiner	*Notropis hudsonius*
Stonecat	*Noturus flavus*
Striped bass (extirpated)	*Morone saxatilis*
Striped shiner	*Luxilis chrysocephalus*
Swallowtail shiner	*Notropis procne*
Tadpole madtom	*Noturus gyrinus*
Tessellated	*Etheostoma olmstedi*
Threespine stickleback	*Gasterosteus aculeatus*
Tonguetied minnow	*Exoglossum laurae*
Trout-perch	*Percopsis omiscomaycus*
Walleye	*Stizostedion vitreum vitreum*
White bass	*Morone chrysops*
White crappie	*Pomoxis annularis*
White perch	*Morone americana*
White sucker	*Catostomus commersoni*
Yellow bullhead	*Ameiurus natalis*
Yellow perch	*Perca flavescens*

Sources: Coon 1999; and Crossman and Van Meter 1979.

and degradation, particularly of wetlands; nutrient pollution; the introduction of bioaccumulative toxic chemicals; and the introduction of nonindigenous species.

Wetlands loss has been significant throughout the Great Lakes region. The states bordering the Great Lakes rank near the top of all states in terms of percent of wetlands lost: Ohio has lost more than 90 percent of its once-extensive wetlands since European settlement. Illinois and Indiana have each lost almost 90 percent; Wisconsin and Michigan have each lost more than 50 percent. Wetlands loss in the basin continues, although at a reduced rate due to protective laws passed in both the United States and Canada, as well as wetlands restoration efforts in specific locales. Shoreline wetlands historically have been lost or degraded for a variety of reasons, including conversion to agricultural or residential land uses, rising water levels, invasion of nonindigenous plant and animal species, and pollution.

Around Lake Ontario's shore, there are 255 wetlands covering more than 17,000 ac (6,880 ha). Most are along the eastern shores of the lake. Forty-three percent of originally present wetlands acres on the Canadian shore of the lake and 60 percent on the United States side have been lost. Wetlands are crucially important habitat for a number of species including birds and fish, particularly in the latter case as spawning or nursery habitat. The stabilization of the lake's water level for operation of the Saint Lawrence Seaway has affected wetlands, particularly at their upper and lower limits. The absence of periodic inundation or drying has affected the wetlands plant community. The lack of water-level variation in coastal wetlands has resulted in the establishment of extensive stands of cattail and domination of other areas by reed canary grass, various shrubs, and the invasive nonindigenous species purple loosestrife. Plant species normally associated with intertidal mudflats have largely disappeared.

Moreover, water level in the lake has increased by about 5 ft (1.5 m) over a century, turning some wetlands into open waters. Along with declines in water quality and invasions of nonindigenous species, the effects of water-level changes have led to gross simplification of some formerly diverse wetland environments. In Cootes Paradise wetland off Hamilton Harbor on the west end of the lake, species diversity of aquatic insects declined dramatically over 40 years, from 57 genera (23 families and 6 orders) in 1948, to nine genera (six families and three orders) in 1978, to only five genera (three families and two orders) in 1995. This change was accompanied by replacement of much of the wetland emergent plant community with open water (Chow-Fraser 1998).

Nutrient pollution, particularly the introduction of excessive phosphorus, has caused excessive algae growth and changed the species composition of the phytoplankton. Lake Ontario has had a shift to more cyanobacteria, which has a lower food value than diatoms for grazers of the phytoplankton. Excessive algae growth in the nearshore areas blocks light penetration and reduces the depth of the photic zone, reducing habitat for aquatic macrophytes. This in turn reduces spawning and nursery habitat for a variety of fishes.

In deeper waters, particularly during summer stratification, increased phyto-plankton densities in the epilimnion reduce the thickness of the photic zone. More important, the increased biomass of algae means a greater mass of decomposing algae in the hypolimnion, and this causes reduced oxygen levels, sometimes low enough that few benthic and demersal organisms can survive.

Overfishing, habitat loss, pollution, and the introduction of nonnative species have dramatically altered the fish assemblages of Lake Ontario. Formerly abundant in the lake but now extinct or relatively scarce are Atlantic salmon, lake trout, lake herring, the deepwater cisco, burbot, the fourhorn sculpin, whitefish, and blue pike. Atlantic salmon, lake trout, and burbot were the most abundant piscivores (fish-eating predators). The deepwater ciscos, largely endemic, are (or were) considered to be a species flock—a group of closely related species that evolved from one or two parent species. These members of the salmonid family (Salmonidae) included the bloater, the deepwater cisco, kiyi, blackfin cisco, and shortnose cisco. Of these the bloater, the kiyi, and the shortnose cisco were reportedly caught in Lake Ontario through the 1960s, but none persisted in the lake after the 1980s.

The whitefish was thought to have disappeared from the lake in the 1970s but has since rebounded in the early 1990s and then declined again. The decline of *Diaporeia* may precipitate a decline in whitefish, as the little amphipod is a major prey species.

Declines in these once-abundant species are due to a number of causes, in some cases acting together to the detriment of the species. The Atlantic salmon migrated up tributary streams to breed; its breeding habitat was degraded or destroyed by sediment pollution, dams, and other barriers to upstream migration. Intense fishing pressure was a major cause of the demise of the deepwater ciscos.

The decline of the lake trout—an emblematic fish of the Great Lakes—was the subject of considerable speculation. Certainly overfishing played a role, with the increasing use of gill nets. But it was the introduction of the sea lamprey that seemed to administer the final blow. The sea lamprey is a parasite. It attaches to fish, uses an abrasive tongue to rasp open a wound, and then sucks blood from the victim until it is satiated or the host dies.

The sea lamprey may have been present originally in Lake Ontario (alone of the Great Lakes, as Niagara Falls presented an insurmountable barrier to this and other marine species), but if it was, its numbers were small. The sea lamprey was first noticed in Lake Ontario in the 1920s, and it began to play an important role in the demise of many of the larger fish species. The sea lamprey is a native of the Atlantic Ocean; it may have been present originally in Lake Ontario or, as some think, it may have entered via the Hudson River and the Erie Canal. Whatever was the case for Lake Ontario, it was not present in the upstream Great Lakes before the construction of the Welland Canal around Niagara Falls. The sea lamprey contributed to the decline of all larger fish in the lake, including not just the lake trout but also burbot, whitefish, and probably the Atlantic salmon as well.

The decline of the larger predatory fish species allowed rapid population growth in prey species. Unfortunately, the native prey species did not benefit as much as some nonindigenous species, in particular the alewife, which together with the rainbow smelt were the dominant fishes in the lake for about a decade.

The lake trout is once again present in Lake Ontario as a result of aggressive stocking programs by lake managers as well as a sea lamprey control program (using strategic applications of a poison). It is questionable, however, whether a self-sustaining population of lake trout can be established in the near future, because of the presence of a particular environmental contaminant. Dioxin is the name given to a family of chlorinated hydrocarbons that were discharged into air and water in the Great Lakes region as a byproduct of some industrial processes, including bleaching of wood fiber. Scientists recently reported that concentrations of a particularly toxic dioxin, though miniscule (around 100 parts per trillion, or about a drop in 500,000 gal/~2 million L of water), were sufficient to cause 100 percent mortality among newly hatched lake trout. So, the decline of the lake trout may have come as a result of reproductive failure, not overharvesting and the sea lamprey, as was thought.

Toxic contamination of the Great Lakes has been a concern for four decades. Particularly with certain persistent toxics (those that do not degrade chemically in the environment, or do so only slowly), the lakes' ecosystems are extremely efficient at accumulating and concentrating those chemicals in the food web. One scientist who studied Lake Ontario said, only half jokingly, that if lake managers really wanted to remove the toxics from the lake, what they would need to do is capture and remove all the fish and put them in a toxic waste landfill. It is in the food web that the highly dispersed environmental contaminants accumulate, and organisms at the top of the food web are likely to carry the largest body concentrations. In Lake Ontario, many of these top predators are birds.

Sources of toxics include municipal sewage treatment plants, industrial effluent, municipal stormwater, and agricultural runoff. Atmospheric deposition onto the Great Lakes is a significant source as well, and Lake Ontario is no exception. In 1988, for example, it was estimated that a third of the DDT and three-fourths of the lead entering the lake got there via the air. Lake Ontario is also in the unfortunate position of being the most downstream of the Great Lakes, so that the Niagara River is a major source of toxics.

Many organisms that are part of the Lake Ontario aquatic food web, particularly birds, suffered population declines as a result of environmental contaminants that underwent bioaccumulation and biomagnification. Fish-eating birds such as Ring-billed Gulls, Great Black-backed Gulls, Night Herons, Herring Gulls, Common Terns, Caspian Terns, and Double-crested Cormorants suffered reproductive failure beginning in the 1970s. High body concentrations of toxic chemicals led to the production of eggs with thin shells that broke easily, elevated rates of embryonic mortality, and deformities such as crossed bills. Bald eagles were similarly affected and their numbers have also been reduced by loss of habitat.

Otter and mink, which also eat lake fish, have also been affected, as have snapping turtles and some fishes, such as (apparently) the lake trout. Even the people who regularly eat fish from the lake seem to be showing some effects: several studies link maternal exposure to lake fish with neurological deficiencies in children. The full ramifications of toxic contamination of Lake Ontario have yet to unfold.

Because of efforts to reduce loadings of toxic chemicals to the Great Lakes, populations of some affected species are slowly recovering. A few, notably the Double-crested Cormorant, have rebounded spectacularly. But the toxic contaminants in some cases are long-lived and will remain in the food web and in sediments for decades. And despite efforts to reduce production and disposal of toxics into the environment, many are still heavily used. For example, in 1992, it was estimated that via atmospheric deposition alone Lake Ontario received about 33,000 lb per year (15,000 kg per year) of pesticides, 93 lb (42 kg) of PCBs, and 110,000 lb per year (50,000 kg per year) of lead, as well as smaller amounts of a large number of toxic chemicals. Lake Ontario will be a sink for persistent toxic chemicals for decades to come.

It is difficult to say what the greatest environmental stressor is in Lake Ontario, but certainly the unrelenting series of introductions of invasive nonindigenous species is a likely candidate. Since the 1800s more than 180 species have been introduced into the Great Lakes. Since Lake Ontario is the downstream lake, species introduced into one of the upstream lakes inevitably make their way down. By broad taxonomic group, the composition of the nonindigenous list is as follows: 61 vascular plants, 33 benthic invertebrates, 26 fishes, 26 phytoplankters, 10 planktonic invertebrates, 10 parasitic invertebrates, three pathogenic bacteria and two pathogenic microsporidians, three pathogenic viruses, three benthic amoebas, two insects, one nekto-benthic invertebrate, one epizootic invertebrate, one parasitic tapeworm, and one parasitic mixosporidian.

Nonindigenous species are introduced accidentally, intentionally by professional resource managers, intentionally by nonprofessionals, and perhaps even maliciously. Not all nonindigenous species that are introduced are problematic: many of Lake Ontario's most prized sportfish are nonindigenous and are stocked annually. Others that may be accidentally introduced do not establish themselves as a successful reproducing population. Some find a niche in the food web without causing problems for many native species. But some have profound impacts that are harmful to native species. The list of these "aquatic nuisance species" is long. Many are from the Caspian Sea, prompting one scientist to exclaim that the Great Lakes are in the midst of an ecological takeover by Caspian species. Some of the most notorious in Lake Ontario are shown in Table 4.10.

The climate of the Great Lakes appears to be changing. Water temperatures have increased slightly over the past century, and a number of migratory birds are arriving sooner and leaving later than they have historically. Air temperatures in recent years have been higher than long-term records. Such changes are consistent with predictions of global warming based on climate models.

Potential impacts based on results of climate models are somewhat speculative, but they include a general loss of water and lowering of lake levels along with

Table 4.10 Notorious Nonindigenous Invaders of Lake Ontario

Species (date of introduction)	Origin	Impacts
Sea lamprey (1835)	Canals	Decline and (in combination with other factors) extinction of native fish species, including the deepwater ciscos (*Coregonus* spp.) as a result of parasitic predation
Alewife (1873)	Bait release by angler	Native fish species decline due to food competition (zooplankton) as well as native fish eggs and juveniles by alewife
Whirling disease (1968)	Unintentional release	High mortality among infected fish
Spiny water flea (1982)	Ballast water	Food web changes and declines in some native and nonindigenous fish and crustacean species through food competition
Eurasian ruffe (1986)	Ballast water	Native fish species declines as a result of competition, esp. yellow perch, several shiner species, trout perch, and brown bullhead
Zebra (1988) and quagga (1989) mussels	Ballast water	Damage to structures in lake; consumption of phytoplankton removes food source for zooplankton and alters food web; toxics are concentrated in mussel feces and enter food web
Round goby (1990)	Ballast water	Native fish species decline as a result of competition for food and predation on juveniles, especially lake trout
Viral hemorrhagic septicemia (2003)	Unknown	High mortality among infected fish; affects a wide range of ecologically and commercially significant fishes

Source: Wisconsin State Environmental Resource Center, http://www.serconline.org/ballast/fact.html.

warmer lake temperatures leading to shifts in fish species composition. While many uncertainties remain, climate change has the potential to further destabilize an ecological system already reacting to multiple stressors.

Lake Ontario's Prospects

It is difficult to be optimistic about the long-term future of Lake Ontario's native species. The system has so many stressors that are difficult if not impossible to control that a return to anything resembling its original state, including its original assemblage of species, is out of the question. Its future will be as a managed lake, with constant and probably increasing interventions by its managers: stocking some species, applying biocides to eliminate other species, regulating pollutants, regulating land use, and regulating water levels—measures that may or may not be successful in restoring balance and health to the Lake Ontario ecosystem. Maintaining the benefits to society from a functioning Lake Ontario ecosystem will be a daunting task.

Conservation Issues of Lakes

The lakes focused on in this chapter represent not only three points along a spectrum of broad climate type (tropical, boreal, and temperate) but also three points along a spectrum of human influence. Lake Victoria is an example of a lake whose ecosystem has been profoundly unbalanced by human actions. Lake Ontario also exemplifies an ecosystem out of balance, distinguished from that of Lake Victoria perhaps only by a matter of degree. Lake Baikal could probably be best characterized as a relatively pristine lake poised at the edge of what is likely to be a long slide into biological disruption and pollution, unless conservation efforts succeed.

For these and all but a relatively few other large lakes in the world, the specifics of environmental decline are depressingly familiar: food web disruption, loss of biodiversity, and degradation of water quality. Behind these are specific human activities or indirect effects of human activities.

The introduction of invasive nonindigenous species is wreaking havoc upon native ecosystems in lakes worldwide. No lake is immune, and few are unaffected. The list of nonindigenous species in Lake Ontario is long but it is likely that it is not unusually so; what distinguishes Lake Ontario is the intensity of scientific study to which it has been subjected. Most lakes are relatively unstudied; at most there might be lists of nonindigenous fish species. But such intense scientific attention is mostly found in lakes in the industrial countries of the world; in the developing countries of Africa, South America, and Southern Asia, it is unusual even for the native species of a lake to be fully identified. Natives undoubtedly have disappeared without a trace. With international travel and trade in raw materials, products, and plants and animals increasing steadily (doubling every 10 to 20 years), it is unlikely that the rate of introduction of nonindigenous species will decline. While this is clearly not good news for native species, some might argue that for people it is not necessarily all bad. Lake Victoria has lost many unique fish species, but the Nile perch that replaced them is much more valuable commercially. Similarly, with good management and good luck, Lake Ontario is capable of producing game fish to support a thriving sport fishing economy. They may not be natives, but most sport fishermen will not complain.

Pollution and degradation of water quality may be more serious in the long term, but as Lake Ontario illustrates, water pollution can (with difficulty) be controlled. Phosphorus loading into the lake has been reduced; primary production (reflected in the density of phytoplankton) has actually been lowered beyond the lake managers' targets as a result of the exploding population of quagga mussels. Toxics are a mixed story: some are relatively easily controlled through regulation; others, like mercury, are highly persistent in the lake environment and cannot easily be removed, even though loading can be reduced.

Finally, climate change resulting from the buildup of greenhouse gases in the atmosphere is discernable throughout the world and will undoubtedly have an effect on lakes worldwide. Rates of precipitation and evaporation will change, affecting water levels, salinity, extent of freeze-over, and the mixing regime. The distribution of species will change, with unpredictable results on lakes' ecosystems.

Further Readings

Books

Burgis, M. J., and P. Morris. 1987. *The Natural History of Lakes*. Cambridge: Cambridge University Press.

Goldschmidt, T. 1996. *Darwin's Dreampond: Drama in Lake Victoria*. Cambridge, MA: MIT Press.

Spring, B. 2001. *The Dynamic Great Lakes*. Baltimore: America House.

Thomson, P. 2007. *Sacred Sea: A Journey to Lake Baikal*. Oxford: Oxford University Press.

Internet Sources

Baikal Web World. n.d. "Destination—Baikal." http://www.bww.irk.ru/index.html.

Environment Canada and U.S. Environmental Protection Agency. "State of the Great Lakes." Reports for 2005 and 2007 available at http://www.epa.gov/glnpo/solec.

Great Lakes Information Network. n.d. "Invasive Species." http://www.great-lakes.net/envt/flora-fauna/invasive/invasive.html.

Great Lakes Information Network. n.d. "Toxic Contamination in the Great Lakes Region." http://www.great-lakes.net/envt/pollution/toxic.html.

Great Lakes Sea Grant Extension Office. n.d. "Great Lakes Water Life Photo Gallery." http://www.glerl.noaa.gov/seagrant/GLWL/GLWLife.html.

National Oceanic and Atmospheric Administration (NOAA). n.d. "Great Lakes Nonindigenous Species List." http://www.glerl.noaa.gov/res/Programs/invasive.

U.S. Environmental Protection Agency. n.d. "Great Lakes Atlas and Sourcebook." http://www.epa.gov/glnpo/atlas/index.html.

U.S. Geological Survey. n.d. "Nonindigenous Aquatic Species." http://nas.er.usgs.gov.

U.S. Geological Survey, Biological Resources Division, and National Aeronautics and Space Administration. n.d. "Land Use History of the United States." http://biology.usgs.gov/luhna/chap6.html.

Appendix

Selected Plants and Animals of Lakes and Reservoirs

Miscellaneous Lake and River Biota

Primary Producers

Reeds	*Phragmites* spp.
Cattails	*Typha* spp.
Water lilies	*Nymphaea* spp.
Pondweed	*Potamogeton* spp.
Coontail	*Ceratophyllum* spp.
Watermilfoil	*Myriophyllum* spp.
Water weeds	*Elodea* spp.

Consumers

Dragonflies	Order Odonata
Mayflies	Order Ephemeroptera
Stoneflies	Order Plecoptera
Caddisflies	Order Trichoptera
Midges (chironomids)	Order Diptera
Green heron	*Butorides virescens*
Waterboatman	*Notonecta* spp.
Water stick	*Ranatra* spp.
Diving beetle	*Dytiscus* spp.

Lake Benthos

Midges (chironomids)	Order Diptera
Nematodes	Order Nematoda
Carp	Family Cyprinidae
Catfish	Order Siluriformes

Lake Pelagic Zone

Primary Producers

Diatoms	Class Bacillariophyceae
Blue-green algae	Phylum Cyanobacteria
Algae	Domain Eukaryota
Dinoflagellates	Phylum Dinoflagellata

Consumers

Waterflea (Daphnia)	Order Cladocera
Copepod	Subclass Copepoda
Alewife	*Alosa pseudoharengus*
Salmon, Trout	Family Salmonidae
Herring, Sardines, Anchovies	Family Clupeidae
Grebe	Order Podicepediformes
Osprey	*Pandion haliaetus*
Tern	Family Sternidae

Salt Lakes

Plankton and Invertebrates

Green alga	*Dunaliella parva*
Red archaeobacteria	Family Halobacteriaceae
Rotifer	*Hexarthra jenkinae, Brachionus plicatilus*
Crustacean	*Parartemia minuta*
Cladocerans	*Moina baylyi, Daphniopsis* spp.
Copepods	*Microcyclops platypus, Microcyclops* spp.
Brine shrimp	*Artemia* spp.
Yabbie crayfish	*Cherax destructor*
Atyid prawn	*Caridina thermophila*
Alkali fly	*Ephedra hyans*
Midges (chironomids)	Order Diptera
Damselflies, dragonflies	Order Odonata

Macrophytes

Sago pondweed	*Potamogeton pectinatus*
Widgeongrass	*Ruppia maritima*
Spiral ditchgrass	*Ruppia occidentalis*
Bulrush	*Scirpus maritimus* var. *paludosus*
Desert saltgrass	*Distichlis stricta*
Nuttal's alkali grass	*Puccinellia nuttalliana*
Chairmaker's bulrush	*Schoenoplectus americanus*
Seaside arrowgrass	*Triglochin maritima*

Fishes

Desert goby	*Chlamydogobius eremius*
Lake Eyre catfish	*Neosilurus* spp.
Dalhousie hardyhead	*Craterocephalus dalhousiensis*
Lake Eyre hardyhead	*Craterocephalus eyresii*
Western mosquitofish	*Gambusia affinis*
Tilapia	Family Cichlidae

Lake Baikal

Freshwater Sponges

No common name	*Lubomirskaja baicalensis*
	Family Lubomirskiidae

Invertebrates

Amphipods (scuds)	Family Gammaridae
Seed shrimp	Class Ostracoda
Segmented worms	Class Oligochaeta
Flatworm	*Baikaloplana valida*
Midge flies (chironomids)	Order Diptera
Epischura (copepod)	*Epischura baicalensis*

Fishes (See Also Tables 4.4 and 4.5)

Baikal omul	*Coregonus autumnalis migratorius*
Golomyanka	*Comephorus baicalensis, C. dybowski*

Mammals

Baikal nerpa	*Phoca sibirica*
Otter	*Lutra lutra*
Sable	*Martes zibellina princeps*

Birds

Mallards	*Anas platyrhynchos*
Common Teal	*Anas crecca*
Northern Pintail	*Anas acuta*
Eurasian Wigeon	*Anas penlope*
Common Goldeneye	*Bucephala clangula*
Bewick's Swan	*Cygnus columbianus bewickii*

Lake Victoria

Primary Producers

Blue-green algae	*Aphanocapsa* spp.
Papyrus	*Cyperus papyrus*
Cattail	*Typha* spp.
Reed	*Phragmites* spp.
Pondweed	*Potamogeton* spp.
Coontail	*Ceratophylum demersum*
Hydrilla	*Hydrilla verticillata*
Knotgrass	*Polygonum* spp.
Hippo grass	*Vossia* spp.
Water hyacinth	*Eichhornia crassipes*

Zooplankton

Copepods	*Thermocyclops neglectus, T. emini; Cyclops* spp., *Diaptomus* spp.
Cladocerans	*Daphnia* spp., *Chydorus* spp., *Leptodora* spp.
Seed shrimp	Class Ostracoda
Rotifers	*Keratella, Asplanchna brightwelli*
Freshwater medusa	*Limnocnida victoriae*
Phantom midges	Family Chaoboridae

Benthos

Lunged snails	*Bulinus* spp., *Biomphalaria* spp.
Common caridina	*Caridina nilotica*

Fishes

Haplochromine cichlids	Family Cichlidae, Genus *Haplochromis*
Tilapiine cichlids	*Oreochromis esculentus, O. variabilis*
Lake Victoria deepwater catfish	*Xenoclarias eupogon*
Ningu	*Labeo victorianus*
Semutundu	*Bagrus docmak*
Nile perch (introduced)	*Lates niloticus*
Nile tilapia (introduced)	*Oreochromis niloticus*
Dagaa (omena)	*Rastrineobola argentea*
Tilapia (introduced)	*Oreochromis leucostictus, Tilapia zillii, T. rendalli*

Birds

Squacco Heron	*Ardeola ralloides*
Grey-headed Gull	*Larus cirrocephalus*

(Continued)

Little Egret	*Egretta garzetta*
Long-tailed Cormorant	*Phalacrocorax africanus*
Papyrus Gonolek	*Laniarius mufumbiri*
Great White Pelican	*Pelecanus onocrotalus*
Pink-backed Pelican	*Pelecanus rufescens*
Blue Swallow	*Hirundo atrocaerulea*
Shoebill	*Balaeniceps rex*
White-winged Black Tern	*Chlidonias leucopterus*
Gull-billed Tern	*Gelochelidon nilotica*
Whiskered Tern	*Chlidonias hybridus*

Mammals

African elephant	*Loxodonta africana*
Black and white colobus monkey	*Colobus guereza adolfi-friederici*
Blue monkey	*Cercopithecus mitis doggetti*
Sitatunga	*Tragelaphus spekii*

Lake Ontario

Fishes (See Also Table 4.9)

Atlantic salmon	*Salmo salar*
Northern pike	*Esox lucius*
Muskellunge	*Esox masquinongy*
Gar	*Lepisosteus oculatus, L. osseus*
Lake trout	*Salvelinus namaycush*
Lake herring	*Coregonus artedii*
Burbot	*Lota lota*
Fourhorn sculpin	*Myoxocephalus quadricornis*
Whitefish	*Coregonus clupeaformis*
Blue pike	*Stizostedion vitreum glaucum*
Bloater	*Coregonus hoyi*
Deepwater cisco	*Coregonus johannae*
Kiyi	*Coregonus kiyi*
Blackfin cisco	*Coregonus nigripinnis*
Shortnose cisco	*Coregonus reighardi*
Sea lamprey (introduced)	*Petromyzon marinus*
Alewife	*Alosa pseudoharengus*
Rainbow smelt	*Osmerus mordax*
Eurasian ruffe (introduced)	*Gymnocephalus cernuus*
Round goby	*Neogobius melanostomus*

Macrophytes (Sheltered Bays)

Tape grass	*Vallisneria americana*
Pondweed	*Potamogeton richardsonii, P. pectinatus,*
	P. gramineus, P. pusillus

Naiad	*Najas flexilis*
Horned pondweed	*Zannichellia palustris*
Water stargrass	*Heteranthera dubia*
Coontail	*Ceratophyllum demersum*
Filamentous algae	*Cladophora* spp.
Muskgrass	*Chara* spp.

Macrophytes (Wetlands)

Waterweed	*Elodea canadensis*
Pondweed	*Potamogeton zosteriformis*
Duckweeds	*Spirodela polyrhiza, Lemna trisulca*
Yellow pond lily	*Nuphar advena*
Fragrant water lily	*Numphaea odorata*
Coontail	*Ceratophyllum demersum*

Macrophytes (Wet Meadows)

Cattail	*Typha angustifolia*
Bluejoint reedgrass	*Calamagrostis canadensis*
Marsh fern	*Thelypteris palustris*
Reed canary grass	*Phalaris arundinacea*
Purple loosestrife	*Lythrum salicaria*

Phytoplankton

| Diatoms | Genera *Cymbella, Stephanodiscus* |
| Green algae | *Phacotus lenticularis, Oocystis* spp., *Staurastrum paradoxum, Ulothrix subconstricta* |

Zooplankton (Epilemnial)

Copepods	*Cyclops bicuspidatus thomasi, Tropocyclops prasinus mexicanus, Eurytemora affinis* (introduced)
Cladocerans	*Bosmina longirostris, Ceriodaphnia lacustris*
Spiny water flea (introduced)	*Bythotrephes cederstroemi*
Fishhook waterflea (introduced)	*Cercopagis pengoi*

Zooplankton (Hypolimnial)

| Copepods | *Diaptomus sicilis, Limnocalanus macrurus* |
| Mysid shrimp | *Mysis relicta* |

Benthos

| Amphipod | *Diporeia hoyi* |
| Annelid worms | Class Oligochaeta; *Stylodrilus heringianus* |

(Continued)

Midge fly	Family Chironomidae
Zebra mussel (introduced)	*Dreissena polymorpha*
Quagga mussel (introduced)	*Dreissena bugensis*

Benthos (Littoral Zone)

Amphipods	*Gammarus* spp.
Annelid worms	Class Oligochaeta; *Stylodrilus heringianus*
Midge fly	Family Chironomidae
Zebra mussel (introduced)	*Dreissena polymorpha*
Dragonflies and damselflies	Order Odonata
Mayflies	Order Ephemeroptera
Dobsonflies and alderflies	Order Megaloptera

Birds

Common Loon	*Gavia immer*
Common Tern	*Sterna hirundo*
Caspian Tern	*Sterna caspia*
Horned Grebe	*Podiceps autirus*
Mallard	*Anas platyrhynchos*
Long-tailed Duck	*Clangula hyemalis*
Double-crested Cormorant	*Phalacrocorax auritus*
Red-breasted Merganser	*Mergus serrator*
Bonaparte's Gull	*Larus philadelphia*
Ring-billed Gull	*Larus delawarensis*
Herring Gull	*Larus argentatus*
Great Black-backed Gull	*Larus marinus*

Glossary

Adaptive Radiation. Diversification of a species as different populations develop adaptations to different ecological niches. Such adaptations could include behavioral and morphological changes, such as the development of specialized mouth parts fitted to a particular type of feeding.

Allochthonous. Material or energy produced from within a system. In the context of aquatic systems, this refers to production within the water body. Compare with autochthonous.

Alluvial Rivers. Rivers whose channels are cut through alluvium, not bedrock. Floodplains are characteristic of alluvial rivers.

Alluvium. Unconsolidated sediment—clay, sand, silt, gravel, cobble, and boulders—deposited by flowing water. Floodplains are composed of alluvium.

Anadromous. Fish that are born and have their initial life stages in freshwater, then migrate to saltwater to live as adults, then return to freshwater to spawn and reproduce. *See also* Catadromous.

Assemblage. The sum total of interacting populations of organisms of a particular type in a particular aquatic habitat. This habitat could be an entire river system or just one particular reach. An assemblage could be one of fishes, macroinvertebrates, or microorganisms. Assemblages of fishes or macroinvertebrates are often used in biomonitoring to assess the biological integrity or health of an aquatic environment. The term assemblage, rather than community, indicates that no assumption is made regarding the persistence and stability of the particular observed grouping of different fish, macroinvertebrates, and other taxa. It may constitute a community linked together in a stable food web, or it may just have been thrown together by happenstance. Much of the literature on biological integrity of rivers assumes that, for a river in a particular ecoregion, there is a predictable assemblage, whether of fish, invertebrates, plankton, or whatever target taxon is used. Any deviation from the assemblage associated with a healthy, undisturbed condition indicates a less-than-healthy river condition.

Astronomical Tide. The periodic, highly predictable rise and fall of water level in large bodies of water due to the gravitational pull of the moon and, to a lesser extent, the sun.

Autochthonous. Matter or energy coming into a system from outside the system. Compare with allochthonous.

Autotrophic. An autotrophic aquatic system (for example, a lake or wetland) is one in which the majority of the food energy comes from plant production in the system. Compare with heterotrophic.

Benthic. Pertaining to or inhabiting the bottom of a body of water.

Benthos. Collective term for all the organisms inhabiting the bottom of a body of water.

Bioaccumulation. The accumulation of pollutants in the body of an organism to a point at which the concentration in the organism is greater than the concentration in the surrounding environmental medium.

Bioconcentration. *See* Bioaccumulation.

Biofilm. A slime layer familiar to all who have slipped on river rocks. It consists of bacteria, fungi, periphyton, and other microscopic organisms together with their secretions, in a kind of matrix that adheres to the substrate.

Biological Diversity (Biodiversity). The variety of all the forms of life in a particular environment. Biodiversity is often expressed in terms of species diversity (how many different species there are) but also encompasses genetic diversity and ecosystem diversity.

Biological Integrity. "The ability of an ecosystem to support and maintain a balanced, integrated, and adaptive community of organisms having a species composition, diversity, and functional organization comparable to the best natural habitats within a region" (Karr and Dudley 1981).

Biomagnification. The cumulative increase in concentration of a persistent (nonbiodegradable) pollutant in organisms at higher trophic levels in the food chain or food web. In aquatic food webs, top predators such as ospreys and seals usually have the highest concentrations.

Catadromous. Migratory fishes that spawn and reproduce in salt water but live as adults in freshwater. *See also* Anadromous.

CPOM (Coarse Particulate Organic Matter). Organic material such as leaves, seeds, and other relatively large plant parts. Particularly in headwater streams in forested regions, CPOM constitutes a major energy source for the lotic ecosystem.

Discharge. The volume of water passing a particular point (or more precisely, passing through a particular cross-section) on a stream or river per unit of time, typically per second. Volume may be reported in either cubic feet or cubic meters. Discharge, therefore, can be given in units of cubic feet per second (typically written as cfs) or cubic meters per second (cumecs).

DOM (Dissolved Organic Material). Carbohydrates, humic acids, and other assorted carbon-based compounds. DOM is derived from biological sources such as leaves and soil organic material.

Ecosystem Engineer. An organism that has a profound effect upon the abiotic (nonliving) environment of an ecosystem, thus creating and maintaining the conditions with which other organisms must cope. Examples include sphagnum mosses, which lower water pH, and beavers, which through their dam-building create pond and wetlands habitat.

Ecotone. A habitat that represents a transition between two distinctly different ecosystems.

Emergent Vegetation. Plants that are rooted underwater but have leaves and stalks that emerge from the water into the air.

Endemic Species. A species that occurs only in one particular region or ecosystem. The region to which a species is endemic may be as small as a particular cave system, or as large as a river system such as the Amazon.

Endorheic Lake. A lake that is the terminus of the internal drainage network of a basin with no outlet to the sea. Endorheic lakes typically form in areas of low precipitation and high evapotranspiration. Some endorheic lakes are highly saline because of millenia of evapotranspiration, for example, the Great Salt Lake of the United States.

Ericaceous Plants. Shrubby plants belonging to the heath family (Ericaceae), which includes such diverse plants as blueberry, snowberry, huckleberry, laurel, rhododendron, and manzanita.

Eutrophic. Nutrient rich.

Eutrophication. A condition in a body of water characterized by abundant plant growth, especially of algae, as a result of enrichment with nutrients, especially phosphorus or nitrogen compounds. Eutrophication may occur naturally or as a result of human activity; the latter case is sometimes referred to as cultural eutrophication.

Flow Regime. The expected pattern of average flows, high flows, and low flows of a stream or river, including their magnitude, frequency, duration, and rate of change. The natural flow regime is seen by some scientists as optimal for the native species inhabiting a stream. Dams, watershed urbanization, and other human activities change a stream's flow regime; over time, changing climate does so too.

FPOM (Fine Particulate Organic Material). This can include pollen, feces from small animals such as zooplankton, and fragments formed as coarse particulate organic matter that is physically broken down in a stream environment.

Headwaters. Those streams in the farthest upstream reaches of a river network; first- or second-order streams that have no tributaries themselves or at most have only first-order tributaries.

Herbaceous Plants. Plants with soft rather than woody tissues; in climate zones with subfreezing temperatures, these plants die back in winter.

Heterotrophic. A heterotrophic aquatic system (for example, a wetland or lake) is one in which the majority of the food energy comes from plant production outside the system, usually from terrestrial sources. Compare with autotrophic.

Hydric Soils. Soils "that formed under conditions of saturation, flooding or ponding long enough during the growing season to develop anaerobic conditions in the upper part" (U.S. Department of Agriculture, Natural Resources Conservation Service). Such soils are often grayish with reddish mottling.

Hydrograph. A graph relating water level (stage) or discharge at a point on a stream or river to time.

Hydroperiod. The periodic or seasonal pattern of changes in water depth in a wetland. Incorporates timing of highs and lows and rate of change of water levels.

Lentic. Pertaining to aquatic habitats in standing or slow-moving water, as in a lake.

Littoral. Pertaining to the shore of a body of water.

Limiting Factor. An environmental factor such as light, oxygen, temperature, or water whose absence or excessive presence limits the growth of population of a particular organism.

Lotic. Pertaining to aquatic habitats in flowing water, as in a stream or river.

Macroinvertebrate. Animals without backbones, such as insect larvae or molluscs, that are large enough to see without the aid of a microscope or magnifying glass.

Macrophyte. A plant large enough to be seen with the unaided eye.

Mesotrophic. Characterized by moderate nutrient levels.

Microbial Loop. Food web or webs in aquatic ecosystems in which production, consumption through several trophic levels, and decomposition take place among microscopic and near-microscopic organisms, such as bacteria, algae, protozoa, and micro-metazoans with relatively weak connections to macro-metazoan food webs. (See Allan 1995 for additional information.)

Mixing Regime. The mixing regime of a lake is the expected annual pattern of vertical mixing of its waters, and includes the extent, frequency, and timing of mixing.

Muck. Decomposed organic material in which the process of decay is so advanced that individual plant parts cannot be identified. Muck is generally black in color and may smell of rotten eggs due to byproducts of decomposition.

Nonpoint Source Pollution. Pollution that enters a body of water during and after a rain event; pollutants are washed off the land surface and travel with the runoff to the receiving body of water. The mix of specific pollutants reflects the prevailing land use and land cover in the watershed.

Oligotrophic. Nutrient poor. Oligotrophic lakes are characterized by little algae, clear water, and high dissolved oxygen levels. Oligotrophic bogs are often dominated by sphagnum moss.

Ombrotrophic Wetlands. Wetlands fed exclusively or mainly by rainwater, which is typically very low in certain key plant nutrients.

Organochlorine. An organic (carbon-based) compound containing one or more atoms of chlorine. Many organochlorine pollutants (for example, DDT and PCBs) are highly persistent in the environment and are subject to biomagnification.

Peat. Partially decomposed and compressed plant material that has accumulated under saturated, anaerobic conditions. Peat may be tannish or reddish in color, or black. When dried it is sometimes used as a fuel because of its high carbon content.

Percolation. The downward movement of water through soil under the influence of gravity.

Peripheral Freshwater Fishes. Fishes from marine families that have taken up residence in fresh waters or that spend part of their lives in fresh and part in saltwater. *See also* Primary freshwater fishes; Secondary freshwater fishes.

Permafrost. A layer of soil that remains frozen year-round.

Pleistocene. A period of the Earth's history characterized by repeated Ice Ages, when glaciers advanced and retreated, beginning almost 2 million years ago and ending with the end of the most recent Ice Age almost 12,000 years ago.

Point Bar. A low sloping ridge of sand and gravel that forms on the inside of meander curves in alluvial streams.

Primary Freshwater Fishes. Fishes that can only live in fresh water. *See also* Peripheral freshwater fishes; Secondary freshwater fishes.

Reach. Any defined length of river channel, from point A to point B.

Redox Reactions. Complementary, reversible chemical reactions in which one or more electrons are transferred from one element or molecule to another. Confusingly, the gain of electrons is referred to as reduction; the loss is referred to as oxidation. Reduction cannot occur without oxidation and vice versa.

Savanna. Tropical or subtropical flat grassland with scattered trees.

Secchi Depth. An estimate of turbidity determined by lowering a special black-and-white disk known as a Secchi disk into the water and recording the depth at which it disappears from view. The greater the Secchi depth, the lower the turbidity, and vice-versa.

Secondary Freshwater Fishes. Fishes that are generally confined to fresh water but can tolerate limited exposure to saltwater, making it possible for them to disperse through estuarial and even nearshore marine waters from one basin to another. *See also* Peripheral freshwater fishes; Primary freshwater fishes.

Seiche. A tide-like rise of water level in a large body of water with corresponding decrease on the other side of the water body, much like the slopping back and forth of water in a basin, caused by winds or atmospheric pressure changes. The periodicity of the seiche is a function of size and shape of the lake or estuary.

Sessile. Rooted in or attached, more or less permanently, to a substrate. Examples of sessile organisms include rooted plants and adult mussels.

Shrub. A woody perennial plant that differs from a tree in that it is smaller and does not have a single trunk, but rather produces several stems from its base.

Speciation. The evolutionary process whereby one or more new species develops from an existing species. An extreme result of speciation is a "species flock," in which a particular environment contains numerous species descended from a single species.

Substrate. The sand, gravel, rock, or organic material of which the bottom of a body of water is made, and to which sessile organisms are attached

Succession. The orderly and predictable process in an ecosystem whereby one community of plants and animals is replaced by another, culminating in a stable "climax" community which exists for a particular set of climatic and geomorphic conditions.

Trophic Status. For a freshwater system, the level or amount of plant growth as measured primarily in terms of algae abundance. Systems with abundant algae and other plant and animal life are referred to as eutrophic. Oligotrophic systems have little plant and animal life. Sometimes the trophic status of a body of water is measured in terms of its concentrations of the key chemical nutrients, phosphorus and nitrogen. Trophic status is also related to turbidity. Oligotrophic lakes have very clear water.

Turbidity. Cloudiness in water caused by sediment or microorganisms in the water column. Turbidity interferes with light penetration into water.

Viscosity. A substance's internal resistance to flow based on intermolecular forces. Molasses has a high viscosity; gasoline has a low viscosity.

Water Column. In a body of water, the volume of water from the surface of the water to the bottom.

Bibliography

Abebe, E., W. Traunspurger, and I. Andrassy, eds. 2006. *Freshwater Nematodes: Ecology and Taxonomy*. Wallingford, UK: CAB International.

Abell, R. A., D. M. Olson, E. Dinerstein, P. T. Hurley, J. T. Diggs, W. Eichbaum, S. Walters, W. Wettengel, T. Allnutt, and C. J. Loucks. 2000. *Freshwater Ecoregions of North America: A Conservation Assessment*. Washington, DC: Island Press.

Albright, T., T. Moorhouse, and T. McNabb. n.d. "The Abundance and Distribution of Water Hyacinth in Lake Victoria and the Kagera River Basin, 1989–2001. Washington, DC: EROS Data Center, U.S. Geological Survey; and Martinize, CA: Clean Lakes, Inc. http://edcintl.cr.usgs.gov/lakespecialfeature.html.

Allan, J. D. 1995. *Stream Ecology: Structure and Function of Running Waters*. London: Chapman and Hall.

Allan, J. D., and A. C. Benke. 2005. "Overview and Prospects." In *Rivers of North America*, ed. A. C. Benke and C. E. Cushing. Burlington, MA: Elsevier Academic Press.

Angermeier, P. L., and J. R. Karr. 1994. "Biological Integrity versus Biological Diversity as Policy Directives." *BioScience* 44: 690–697.

Appalachian Power Company. 2006. Claytor Hydroelectric Project. FERC Project No. 739. Integrated License Process Pre-Application Document. Roanoke, VA: Appalachian Power Company.

Barnes, K. K., D. W. Kolpin, M. T. Meyer, E. M. Thurman, E. T. Furlong, S. D. Zaugg, and L. B. Barber. 2002. "Water-Quality Data for Pharmaceuticals, Hormones, and Other Organic Wastewater Contaminants in U.S. Streams, 1999–2000." Open-File Report 02-94. Reston, VA: U.S. Geological Survey.

Bayley, P. B. 1995. "Understanding Large River-Floodplain Ecosystems." *BioScience* 45 (3): 153–158.

215

Behangana, M., and J. Arusi. 2004. "The Distribution and Diversity of Amphibian Fauna of Lake Nabugabo and Surrounding Areas." *African Journal of Ecology* 42 (Suppl. 1): 6–13.

Black, S. H., and M. Vaughan. 2006. "Endangered Insects." In *The Encyclopedia of Insects*, ed. V. Resh and R. Carde. San Diego, CA: Academic Press. http://www.xerces.org/articles.htm.

Bogatov, V., S. Sirotsky, and D. Yuriev. 1995. "The Ecosystem of the Amur River." In *River and Stream Ecosystems,* ed. C. E. Cushing, K. W. Cummins, and G. W. Minshall, 601–613. Ecosystems of the World, 22. Amsterdam: Elsevier.

Bogutskaya, N., A. Naseka, and N. E. Mandrak. 2004. "Threats to the Fishes of Khanka Lake, a Large, Shallow Asian Lake." Abstract published in Great Lakes Need Great Watersheds, *Proceedings of the Annual Conference of the International Association for Great Lakes,* p. 13. (Retrieved from Water Resources Abstract database.)

Bolen, E. G., L. M. Smith, and H. L. Schramm, Jr. 1989. "Playa Lakes: Prairie Wetlands of the Southern High Plains." *BioScience* 39 (9): 615–624.

Bondarenko, N. A., A. Tuji, and M. Nakanishi. 2006. "A Comparison of Phytoplankton Communities between the Ancient Lakes Biwa and Baikal." *Hydrobiologia* 568 (S): 25–29.

Bonetta, A. A., and I. R. Wais. 1995. "Southern South American Streams and Rivers." In *River and Stream Ecosystems*, ed. C. E. Cushing, K. W. Cummins, and G. W. Minshall, 257–294. Ecosystems of the World, 22. Amsterdam: Elsevier.

Botch, M. S., and V. V. Masing. 1983. "Mire Ecosystems in the U.S.S.R." In *Mires: Swamp, Bog, Fen, and Moor—Regional Studies*, ed. A. J. P. Gore, 95–152. Ecosystems of the World, 4B. Amsterdam: Elsevier.

Boulton, A. J., S. Findlay, P. Marmonier, E. H. Stanley, and H. M. Valett. 1998. "The Functional Significance of the Hyporheic Zone in Streams and Rivers." *Annual Review of Ecology and Systematics* 29: 59–81.

Brinson, M. M. 1993. "A Hydrogeomorphic Classification for Wetlands." Technical Report WRP-DE-4, NTIS No. AD A270 053. Vicksburg, MS: U.S. Army Engineer Waterways Experiment Station.

Brittain, J. E., and T. J. Eikeland. 1988. "Invertebrate Drift—a Review." *Hydrobiologia* 166 (1): 77–93.

Brix, H., B. K. Sorrell, and P. T. Orr. 1992. "Internal Pressurization and Convective Gas Flow in Some Emergent Freshwater Macrophytes." *Limnology and Oceanography* 37 (7): 1420–1433.

Bronmark, C., and L.-A. Hansson. 2005. *The Biology of Lakes and Ponds.* 2nd ed. Biology of Habitats Series. New York: Oxford University Press.

Browder, J. O., M. A. Pedlowski, and P. M. Summers. 2004. "Land Use Patterns in the Brazilian Amazon: Comparative Farm-Level Evidence from Rondonia." *Human Ecology: An Interdisciplinary Journal* 32 (2): 197–225.

Brunello, A. J., V. C. Molotov, B. Dugherkhuu, C. Goldman, E. Khamaganova, T. Strijhova, and R. Sigman. 2006. *Lake Baikal: Experience and Lessons Learned Brief.* South Lake Tahoe, CA: Tahoe-Baikal Institute.

Brunello, A. J., V. C. Molotov, B. Dugherkhuu, C. Goldman, E. Khamaganova, T. Strijhova, and R. Sigman. 2004. *Lake Baikal Watershed: Lake Basin Management Experience Brief.* South Lake Tahoe, CA: Tahoe-Baikal Institute.

Bugenyi, F. N. B. 2004. "The Problem of Rehabilitating Lakes and Wetlands in Developing Countries: The Case Example of East Africa." In *The Lakes Handbook,* ed. P. E.

O'Sullivan and C. S. Reynolds, 523–533. *Volume 2, Lake Restoration and Rehabilitation.* Malden, MA: Blackwell.

Burgis, M. J., and P. Morris. 1987. *The Natural History of Lakes.* Cambridge: Cambridge University Press.

Campbell, D. 2005. *A Land of Ghosts: The Braided Lives of People and the Forest in Far Western Amazonia.* Boston: Houghton Mifflin.

Carolsfeld, J., B. Harvey, C. Ross, and A. Baer, eds. 2003. *Migratory Fishes of South America. Biology, Fisheries and Conservation Status.* Washington, DC: The International Bank for Reconstruction and Development and the World Bank (International Development Research Center and World Fisheries Trust). http://www.idrc.ca/openebooks/114-0.

Cherniak, M. 1989. *The Hawk's Nest Incident: America's Worst Industrial Disaster.* New Haven, CT: Yale University Press (reprint edition).

Chow-Fraser, P., V. Lougheed, V. Le Thiec, B. Crosbie, L. Simser, and J. Lord. 1998. "Long-Term Response of the Biotic Community to Fluctuating Water Levels and Changes in Water Quality in Cootes Paradise Marsh, a Degraded Coastal Wetland of Lake Ontario." *Wetlands Ecology and Management* 6 (1): 19–42.

Colburn, E. A. 2004. *Vernal Pools: Natural History and Conservation.* Blacksburg, VA: McDonald and Woodward.

Coon, T. G. 1999. "Ichthyofauna of the Great Lakes Basin." In *Great Lakes Fisheries Policy and Management,* ed. W. W. Taylor and C. P. Ferreri, 55–72. East Lansing: Michigan State University Press.

Cowardin, L. M., F. C. Golet, and E. T. LaRoe. 1979. *Classification of Wetlands and Deepwater Habitats of the United States.* Office of Biological Services, FWS/OBS-79/31. Washington, DC: U.S. Department of the Interior, U.S. Fish and Wildlife Service. http://www.npwrc.usgs.gov/resource/wetlands/classwet/index.htm#contents.

Crane, K., B. Hecker, and V. Golubev. 1991. "Heat Flow and Hydrothermal Vents in Lake Baikal, U.S.S.R." *EOS Eostaj* 72 (52): 585, 588–589.

Cross, D. H., compiler. 1993. *Waterfowl Management Handbook.* Fish and Wildlife Leaflet 13, a compilation of U.S. Fish and Wildlife Leaflets published 1988–1993 about the management of waterfowl and their habitat. Lafayette, LA: U.S. Fish & Wildlife Service (National Wetlands Research Center). http://www.nwrc.usgs.gov/wdb/pub/wmh/contents.html.

Crossman, E. J., and H. D. Van Meter. 1979. "Annotated List of the Fishes of the Lake Ontario Watershed." Technical Report No. 36, June. Ann Arbor, MI: Great Lakes Fishery Commission.

Crul, R. C. M. 1995. *Limnology and hydrology of Lake Victoria.* Paris: UNESCO Publishing.

Cummins, K. W., C. E. Cushing, and G. W. Minshall. 1995. "An Overview of Stream Ecosystems." In *River and Stream Ecosystems,* ed. C. E. Cushing, K. W. Cummins, and G. W. Minshall, 1–8. Ecosystems of the World, 22. Amsterdam: Elsevier.

Cushing, C. E., and J. D. Allan. 2001. *Streams: Their Ecology and Life.* San Diego: Academic Press.

Czaya, E. 1981. *Rivers of the World.* New York: Van Nostrand Reinhold.

Dahl, T. E. 2006. "Status and Trends of Wetlands in the Conterminous United States 1998 to 2004." Washington, DC: U.S. Department of the Interior; Fish and Wildlife Service. http://wetlandsfws.er.usgs.gov/status_trends/national_reports/trends_2005_report.pdf.

Darlington, P. J., Jr. 1982. *Zoogeography.* Maliber, FL: Krieger Publishing.

Davies, B. R., and K. F. Walker. 1986. *The Ecology of River Systems.* Dordrecht: Dr. W. Junk Publishers (Kluwer).

Davis, C. A., and L. M. Smith. 1998. "Ecology and Management of Migrant Shorebirds in the Playa Lakes Region of Texas." *Wildlife Monographs* 140: 1–45.

Duggan, I. C., S. A. Bailey, R. I. Colautti, D. K. Gray, J. C. Makarewics, and H. J. MacIsaac. 2003. "Biological Invasions in Lake Ontario: Past, Present and Future." In *State of Lake Ontario (SOLO)—Past, Present and Future,* ed. M. Munawar, 1–17. Ecovision World Monograph Series. Burlington, Ontario, Canada: Aquatic Ecosystem Health and Management Society.

Duker, L., and L. Borre. 2001. *Biodiversity Conservation of the World's Lakes: A Preliminary Framework for Identifying Priorities.* LakeNet Report Series, Number 2. Annapolis, MD: Monitor International.

Environment Canada. 2005. "A National Ecological Framework for Canada." Hull, Quebec: Environment Canada. http://www2.ec.gc.ca/soer-ree/English/Framework/default.cfm.

Euliss, N. H., Jr., D. M. Mushet, and D. A. Wrubleski. 1999. "Wetlands of the Prairie Pothole Region: Invertebrate Species Composition, Ecology, and Management." In *Invertebrates in Freshwater Wetlands of North America: Ecology and Management,* ed. D. P. Batzer, R. B. Rader, and S. A. Wissinger, 471–514. New York: Wiley. Northern Prairie Wildlife Research Center (Jamestown, ND) http://www.npwrc.usgs.gov/resource/wetlands/pothole/index.htm (Version 02SEP99).

Federal Interagency Committee for Wetland Delineation. 1989. *Federal Manual for Identifying and Delineating Jurisdictional Wetlands.* Washington, DC: U.S. Army Corps of Engineers, U.S. Environmental Protection Agency, U.S. Fish and Wildlife Service, and USDA Soil Conservation Service (Cooperative technical publication).

Field, R. T., and K. R. Philipp. 2000. "Vegetation Changes in the Freshwater Tidal Marsh of the Delaware Estuary." *Wetlands Ecology and Management* 8: 79–88.

Finlayson, C. M., N. C. Davidson, A. G. Spiers, and N. J. Stevenson. 1999. "Global Wetland Inventory—Current Status and Future Priorities." *Marine Freshwater Research* 50 (8): 717–727.

Flessa, K.W. 2004. *Ecosystem Services and the Value of Water in the Colorado River Delta and Estuary, USA and Mexico: Guidelines for Mitigation and Restoration,* 79–86. International Seminar on Restoration of Damaged Lagoon Environments, Matsue, Japan.

Fraser, L. H., and P. A. Keddy, eds. 2005. *The World's Largest Wetlands. Ecology and Conservation.* Cambridge: Cambridge University Press.

Garmaeva, T. 2001. "Lake Baikal: Model for Sustainable Development of the Territory. Case Report." *Lakes & Reservoirs: Research and Management* 6: 253–257.

Gerritsen, J., and M. T. Barbour. 2000. "Apples, Oranges, and Ecoregions: On Determining Pattern in Aquatic Assemblages." *Journal of the North American Benthological Society* 19 (3): 487–496.

Giller, P. S., and B. Malmqvist. 1998. *The Biology of Streams and Rivers.* Oxford: Oxford University Press.

Godoy, J. R., G. Petts, and J. Salo. 1999. "Riparian Flooded Forests of the Orinoco and Amazon Basins: A Comparative Review." *Biodiversity and Conservation* 8 (4): 551–586.

Goldschmidt, T., F. Witte, and J. Wanink. 1993. "Cascading Effects of the Introduced Nile Perch on the Detritivorous/Phytoplanktivorous Species in the Sublittoral Areas of Lake Victoria." *Conservation Biology* 7 (3): 686–700.

Goldschmidt, T. 1996. *Darwin's Dreampond: Drama in Lake Victoria.* Cambridge, MA: MIT Press (transl. by S. Marx-McDonald).

Google Earth. n.d. Indispensable resource for viewing large ecosystems and landforms. http://earth.google.com/.

Gorthner, A. 1994. "What Is an Ancient Lake?" *Ergebnisse der Limnologie* 44: 97.

Harris, M. B., W. Tomas, G. Mourao, C. J. da Silva, E. Guimaraes, F. Sonodas, and E. Fachim. 2005. "Safeguarding the Pantanal Wetlands: Threats and Conservation Initiatives." *Conservation Biology* 19 (3): 714–720, June.

Haukos, D. A., and L. M. Smith. 1992. "Ecology of Playa Lakes." In *Waterfowl Management Handbook*, Section 13.3.7. Washington, DC: U.S. Department of the Interior, Fish and Wildlife Service.

Heino, J., T. Muotka, R. Paavola, H. Hamalainen, and E. Koskenniemi. 2002. "Correspondence Between Regional Delineations and Spatial Patterns in Macroinvertebrate Assemblages of Boreal Headwater Streams." *Journal of the North American Benthological Society* 21 (3): 397–413.

Helfman, G. S., B. B. Collette, and D. E. Facey. 1997. *The Diversity of Fishes.* Malden, MA: Blackwell Science.

Hershey, A. E., J. Pastor, B. J. Peterson, and G. W. Kling. 1993. "Stable Isotopes Resolve the Drift Paradox for *Baetis* Mayflies in an Arctic River." *Ecology* 74 (8): 2315–2325.

Hoagland, B. 2000. *The Vegetation of Oklahoma: A Classification for Landscape Mapping and Conservation Planning.* Norman: Oklahoma Natural Heritage Inventory and Department of Geography, University of Oklahoma.

Hoagland, B. W., and S. L. Collins. 1997. "Heterogeneity in Shortgrass Prairie Vegetation: The Role of Playa Lakes." *Journal of Vegetation Science* 8 (2): 277–286.

Holland, R. F., and S. K. Kain. 1981. "Insular Biogeography of Vernal Pools in the Central Valley of California." *American Naturalist* 117 (1): 24–37.

Horner, R. R., A. L. Azous, K. O. Richter, S. S. Cooke, L. E. Reinelt, and K. Ewing. 1997. "Wetlands and Stormwater Management Guidelines." In *Wetlands and Urbanization: Implications for the Future*, ed. A. L. Azous and R. R. Horner, 299–323. Boca Raton, FL: Lewis Publishers.

Hughes, N. F. 1986. "Changes in the Feeding Biology of the Nile Perch, *Lates niloticus* (L.) (Pisces: Centropomidae), in Lake Victoria, East Africa Since Its Introduction in 1960, and Its Impact on the Native Fish Community of the Nyanza Gulf." *Journal of Fish Biology* 29 (5): 541.

Hughes, R. M., E. Rexstad, and C. E. Bond. 1987. "The Relationship of Aquatic Ecoregions, River Basins and Physiographic Provinces to the Ichthyogeographic Regions of Oregon." *Copeia* (2): 423–432.

Hynes, H. B. N. 1979. *The Ecology of Running Waters.* Liverpool, UK: Liverpool University Press.

Ingold, C. T. 1979. "Advances in the Study of So-Called Aquatic Hyphomycetes." *American Journal of Botany* 66 (2): 218–226.

Iwakuma, T., T. Inoue, T. Kohyama, M. Osaki, H. Simbolon, H. Tachibana, H. Takahashi, N. Tanaka, and K. Yabe, eds. 2000. *Proceedings of the International Symposium on Tropical Peat Lands, Bogor, Indonesia, November 22–24, 1999.* Sapporo, Japan: Hokkaido University Graduate School of Environmental Earth Science.

Janauer, G., and M. Dokulil. 2006. "Macrophytes and Algae in Running Waters." In *Biological Monitoring of Rivers*, ed. G. Ziglio, M. Siligardi, and G. Flaim, 90–109. New York: John Wiley and Sons.

Jellison, R., H. Adams, and J. M. Melack. 2001. "Re-appearance of Rotifers in Hypersaline Mono Lake, California, During a Period of Rising Lake Levels and Decreasing Salinity." *Hydrobiologia* 466: 39–43.

Jenkins, R. E., and N. Burkhead. 1994. *Freshwater Fishes of Virginia.* Bethesda, MD: American Fisheries Society.

Junk, W. J. 1986. "Aquatic Plants of the Amazon System." In *Ecology of River Systems,* ed. B. R. Davies and K. F. Walker, 319–338. Dordrecht: Dr. W. Junk Publishers (Kluwer).

Junk, W. J., and M. T. F. Piedade. 1997. "Plant Life in the Floodplain with Special Reference to the Herbaceous Plants." In *The Central Amazon Floodplain,* ed. W. J. Junk, 147–181. Berlin: Springer.

Kaluzhnayal, O. V., S. I. Belikov, H. C. Schröder, M. Rothenberger, S. Zapf, J. A. Kaandorp, A. Borejko, I. M. Müller, and W. E. G. Müller. 2005. "Dynamics of Skeleton Formation in the Lake Baikal Sponge *Lubomirskia baicalensis.*" Part I. Biological and biochemical studies. *Naturwissenschaften* 92 (3): 128–133.

Karr, J. R., and D. R. Dudley. 1981. "Ecological Perspective on Water Quality Goals." *Environmental Management* 5: 55–68.

Kayombo, S., and S. E. Jorgensen. 2006. "Lake Victoria: Experience and Lessons Learned Brief." Kusatsu-shi, Japan: International Lake Environment Committee Foundation for Sustainable Management of World Lakes and Reservoirs. http://www.ilec.or.jp/eg/lbmi/reports/27_Lake_Victoria_27February2006.pdf.

Kunisue, T., T. B. Minh, K. Fukuda, M. Watanabe, S. Tanabe, and A. M. Titenko. 2002. "Seasonal Variation of Persistent Organochlorine Accumulation in Birds from Lake Baikal, Russia, and the Role of the South Asian Region as a Source of Pollution for Wintering Migrants." *Environmental Science and Technology* 36: 1396–1404.

Larsen, D. P., J. M. Omernik, R. M. Hughes, C. M. Rohm, T. R., Whitier, A. J. Kinney, A. L. Gallant, and D. R. Dudley. 1986. "Correspondence Between Spatial Patterns in Fish Assemblages in Ohio Streams and Aquatic Ecoregions." *Environmental Management* 10 (6): 815–828.

Lewis, W. M., Jr., S. K. Hamilton, and J. F. Saunders III. 1995. "Rivers of Northern South America." In *River and Stream Ecosystems,* ed. C. E. Cushing, K. W. Cummins, and G. W. Minshall, 219–256. Ecosystems of the World, 22. Amsterdam: Elsevier.

Lung'Ayia, H. B. O., A. M'Harzi, M. Tackx, J. Gichuki, and J. J. Symoens. 2000. "Phytoplankton Community Structure and Environment in the Kenyan Waters of Lake Victoria." *Freshwater Biology* 43: 520–543.

Luoto, M., R. K. Heikkinen, and T. R. Carter. 2004. "Loss of Palsa Mires in Europe and Biological Consequences." *Environmental Conservation* 31: 30–37.

Lytle, D. A., and N. L. Poff. 2004. "Adaptation to Natural Flow Regimes." *TRENDS in Ecology and Evolution* 19 (2): 94–100.

Marques, M. I., J. Adis, G. B. dos Santos, and L. D. Battirola. 2006. "Terrestrial Arthropods from Tree Canopies in the Pantanal of Mato Grosso, Brazil." *Revista Brasileira de Entomologia* 50 (2): 257–267.

Mateus, A. F., J. M. F. Penha, and M. Petrere. 2004. "Fishing Resources in the Rio Cuiabá Basin, Pantanal do Mato Grosso, Brazil." *Neotropical Ichthyology* 2 (4): 217–227.

Maxwell, J. R., C. J. Edwards, M. E. Jensen, S. J. Paustian, H. Parrott, and D. M. Hill. 1995. *A Hierarchical Framework of Aquatic Ecological Units in North America (Nearctic Zone).*

Gen. Tech. Rep. NC-176. St. Paul, MN: U.S. Department of Agriculture, Forest Service, North Central Forest Experiment Station.

McKnight, D. M., D. K. Niyogi, A. S. Alger, A. Bomblies, P. A. Conovitz, and C. M. Tate. 1999. "Dry Valley Streams in Antarctica: Ecosystems Waiting for Water." *BioScience* 49 (12): 985–995.

McMahon, G., S. M. Gregonis, S. W. Waltman, J. M. Omernik, T. D. Thorson, J. A. Freeouf, A. H. Rorick, and J. E. Keyes. 2001. "Developing a Spatial Framework of Common Ecological Regions for the Coterminous United States." *Environmental Management* 28 (3): 293–316.

Meade, R. H., J. M. Rayol, S. C. Da Conceicao, and J. R. G. Natividade. 1991. "Backwater Effects in the Amazon River Basin of Brazil." *Environmental Geology* 18 (2): 105–114.

Mel'nikov, Yu. I. 2006. "The Migration Routes of Waterfowl and Their Protection in Baikal Siberia." In *Waterbirds Around the World,* ed. G. C. Boere, C. A. Galbraith, and D. A. Stroud, 357–362. Edinburgh: The Stationery Office.

Middleton, B. A. 2002. *Flood Pulsing in Wetlands: Restoring the Natural Hydrological Balance.* New York: John Wiley and Sons, Inc.

Mills, E. L., J. M. Casselman, R. Dermott, J. D. Fitzsimons, G. Gal, K. T. Holeck, J. A. Hoyle, O. E. Johannsson, B. F. Lantry, J. C. Makarewicz, et al. 2005. *A Synthesis of Ecological and Fish-Community Changes in Lake Ontario, 1970–2000.* Ann Arbor, MI: Great Lakes Fishery Commission.

Mitsch, W. J., and J. G. Gosselink. 2000. *Wetlands.* New York: John Wiley and Sons, Inc.

Mizandrontsev, I. B., and K. N. Mizandrontseva. 2001. "The Oxygen Cycle in Lake Baikal." *Water Resources* 28 (5): 502–508. Translated from *Vodnye Resursy* 28 (5): 552–558.

Moorhead, D. L., D. L. Hall, and M. R. Willig. 1998. "Succession of Macroinvertebrates in Playas of the Southern High Plains, USA." *Journal of the North American Benthological Society* 17 (4): 430–442.

Muli, J. R., and K. M. Mavuti. 2001. "The Benthic Macrofauna Community of Kenyan Waters of Lake Victoria." *Hydrobiologia* 458: 83–90.

NRC (National Research Council). 1992. Committee on Restoration of Aquatic Ecosystems: Science, Technology, and Public Policy, Water Science and Technology Board, Commission on Geosciences, Environment, and Resources, National Research Council. *Restoration of Aquatic Ecosystems Science, Technology, and Public Policy.* Washington, DC: National Academy Press.

NRC (National Research Council). 1995. Committee on Characterization of Wetlands, Water Science and Technology Board, Board on Environmental Studies and Toxicology, Commission on Geosciences, Environment, and Resources, National Research Council. *Wetlands: Characteristics and Boundaries.* Washington, DC: National Academy Press.

Odada, E. O., D. O. Olago, K. Kulindwa, M. Ntiba, and S. Wandiga. 2004. "Mitigation of Environmental Problems in Lake Victoria, East Africa: Causal Chain and Policy Options Analyses." *AMBIO: A Journal of the Human Environment* 33 (1): 13–23.

Odum, W. E., T. J. Smith, III, J. K. Hoover, and C. C. McIvor. 1984. "The Ecology of Tidal Freshwater Marshes of the United States East Coast: A Community Profile." FWS/OBS-83/17. Washington, DC: U.S. Department of the Interior.

Omernik, J. M. 1987. "Ecoregions of the Conterminous United States." *Annals of the Association of American Geographers* 77: 118–125.

Omernik, J. M., and R. G. Bailey. 1997. "Distinguishing Between Watersheds and Ecoregions." *Journal of the American Water Resources Association* 33 (5): 935–949.

O'Sullivan, P. E., and C. S. Reynolds, eds. 2004. *The Lakes Handbook*, 523–533. *Volume 2, Lake Restoration and Rehabilitation*. Malden, MA: Blackwell.

Pan, Y., R. J. Stevenson, B. H. Hill, and A. T. Herlihy. 2000. "Ecoregions and Benthic Diatom Assemblages in Mid-Atlantic Highlands Streams, USA." *Journal of the North American Benthological Society* 19 (3): 518–540.

Pile, A. J., M. R. Patterson, M. Savarese, V. I. Chernykh, and V. A. Fialkov. 1997. "Trophic Effects of Sponge Feeding Within Lake Baikal's Littoral Zone. 2. Sponge Abundance, Diet, Feeding Efficiency, and Carbon Flux." *Limnology and Oceanography* 42 (1): 178–184.

Pimentel, D., L. Lach, R. Zuniga, and D. Morrison. 2000. "Environmental and Economic Costs of Nonindigenous Species in the United States." *BioScience* 50 (1): 53–65.

Pimentel, D., R. Zuniga, and D. Morrison. 2005. "Update on the Environmental and Economic Costs Associated with Alien-Invasive Species in the United States." *Ecological Economics* 52 (3): 273–288.

de Pinho, J. B., L. E. Lopes, D. H. de Moraisi, and A. M. Fernandes. 2006. "Life History of the Mato Grosso Antbird Cercomacra melanaria in the Brazilian Pantanal." *Ibis* 148 (2): 321–329.

Por, F. D. 1995. *The Pantanal of Mato Grosso (Brazil): World's Largest Wetlands. Volume 73, Monographiae Biologicae.* Dordrecht: Kluwer Academic Publishers.

Postel, S., and B. Richter. 2003. *Rivers for Life: Managing Water for People and Nature.* Washington, DC: Island Press.

Pott, A., and V. J. Pott. 2004. "Features and Conservation of the Brazilian Pantanal Wetland." *Wetlands Ecology and Management* 12: 547–552.

Quinn, F. H. 1992. "Hydraulic Residence Times for the Laurentian Great Lakes." *Journal of Great Lakes Research* 18 (1): 22–28.

Reschke, C. 1990. *Ecological Communities of New York State.* New York Natural Heritage Program. Latham: New York State Department of Environmental Conservation.

Ricciardi, A., R. J. Neves, and J. B. Rasmussen. 1998. "Impending Extinctions of North American Freshwater Mussels (Unionoida) Following the Zebra Mussel (Dreissena polymorpha) Invasion." *Journal of Animal Ecology* 67 (4): 613–619.

Rossiter, A., and H. Kawanabe, eds. 2000. *Ancient Lakes: Biodiversity, Ecology and Evolution. Volume 31, Advances in Ecological Research.* London: Academic Press.

Ryder, R. A., and J. A. Orendorff. 1999. "Embracing Biodiversity in the Great Lakes Ecosystem." In *Great Lakes Fisheries Policy and Management,* ed. W. W. Taylor and C. P. Ferreri, 113–143. East Lansing: Michigan State University Press.

Sculthorpe, C. D. 1967. *The Biology of Aquatic Vascular Plants.* London: Arnold.

Seabloom, E. W., and A. G. van der Valk. 2003. "The Development of Vegetative Zonation Patterns in Restored Prairie Pothole Wetlands." *Journal of Applied Ecology* 40: 92–100.

Shimaraev, M. N., N. G. Granin, V. M. Domysheva, A. A. Zhdanov, L. P. Golobokova, R. Y. Gnatovskii, V. V. Tsekhanovskii, and V. V. Blinov. 2003. "Water Exchange Between Bed Depressions in Baikal." *Water Resources* 30 (6): 623–626. Translated from *Vodnye Resursy* 30 (6): 678–681.

Sideleva, V. G. 2000. "The Ichthyfauna of Lake Baikal, with Special Reference to Its Zoogeographical Relations." In *Ancient Lakes: Biodiversity, Ecology and Evolution,* ed. A.

Rossiter and H. Kawanabe, 81–96. *Volume 31, Advances in Ecological Research.* London: Academic Press.

Sparks, R. E. 1995. "Need for Ecosystem Management of Large Rivers and Their Flood-plains." *Bioscience* 45 (3): 168–182.

Stanford, J. A., and J. V. Ward. 1988. "The Hyporheic Habitat of River Ecosystems (letter)." *Nature* 335 (September 1): 64–66.

Sternberg, H. O. 1987. "Aggravation of Floods in the Amazon River as a Consequence of Deforestation?" *Geografiska Annaler Series A, Physical Geography* 69 (1): 201–219.

Strahler, A. N. 1952. "Dynamic Basis of Geomorphology." *Geological Society of America Bulletin* 63: 923–938.

Swarts, F. 2000. "The Pantanal in the 21st Century: For the Planet's Largest Wetland, an Uncertain Future." In *The Pantanal of Brazil, Bolivia and Paraguay.* Waterland Research Institute. Gouldsboro, PA: Hudson MacArthur. http://www.pantanal.org.

Tahoe-Baikal Institute. 2004. "Baikal Reader." South Lake Tahoe, CA: Tahoe-Baikal Institute. www.tahoebaikal.org.

Taylor, W. W., and C. P. Ferreri, eds. 1999. *Great Lakes Fisheries Policy and Management.* East Lansing: Michigan State University Press.

Teodoru, C., D. F. McGinnis, A. Wuest, and B. Wehrli. 2006. "Nutrient Retention in the Danube's Iron Gate Reservoir." *Eos* (Transactions of the American Geophysical Union) 87 (38): 385.

Theiling, C. H., D. Wilcox, C. Korschgen, H. DeHaan, T. Fox, J. Rhoweder, L. Robinson, H. Stoerker, J. Janvrin, and S. Whitney. 2000. "Habitat Needs Assessment for the Upper Mississippi River System: Summary Report." La Crosse, WI: U.S. Geological Survey. USGS Upper Midwest Environmental Sciences Center. http://www.umesc.usgs.gov/habitat_needs_assessment/summ_report.html.

Thieme, M. L., R. Abell, M. L. J. S. Stiassny, P. Skelton, N. Burgess, B. Lehner, E. Dinerstein, D. Olson, G. Teugels, and A. Kamdem-Toham. 2005. *Freshwater Ecoregions of Africa and Madagascar: A Conservation Assessment.* Washington, DC: Island Press.

Tiner, R. W. 1997. "Technical Aspects of Wetlands: Wetland Definitions and Classifications in the United States." In *National Water Summary on Wetland Resources.* Water Supply Paper 2425. Reston, VA: U.S. Geological Survey. http://water.usgs.gov/nwsum/WSP2425.

Trayler, K. 2000. "River Restoration Manual 7. Stream Ecology." East Perth: Water and Rivers Commission. http://www.wrc.wa.gov.au/public/RiverRestoration/publica tions/rr7/part2.pdf.

Treece, M. W., Jr. 2003. "Seasonal and Spatial Variability of Pesticides in Streams of the Upper Tennessee River Basin, 1996–1999." Water-Resources Investigations Report 03-4006. Reston, VA: U.S. Geological Survey. http://pubs.usgs.gov/wri/wri034006/PDF/wrir034006part1.pdf#search=%22flora%20fauna%20copper%20creek%20clinch%22.

Trolle, M. 2003. "Mammal Survey in the Southeastern Pantanal, Brazil." *Biodiversity and Conservation* 12: 823–836.

University of Wisconsin Sea Grant Institute. n.d. "Great Lakes Online, Gifts of the Glaciers." http://www.seagrant.wisc.edu/communications/greatlakes/GlacialGift.

U.S. Environmental Protection Agency, Great Lakes National Program Office. 1995. *The Great Lakes: An Environmental Atlas and Sourcebook,* 3rd ed. Chicago: U.S. Environmental Protection Agency, Great Lakes National Program Office. Produced jointly with the

Government of Canada. http://www.epa.gov/glnpo/atlas/glat-ch2.html (updated March 9, 2006).

U.S. Environmental Protection Agency, Office on Wetlands, Oceans, and Watersheds. 1997. *Volunteer Stream Monitoring. A Methods Manual.* EPA 841-B-97-003. Washington, DC: U.S. EPA.

U.S. Geological Survey, U.S. Department of the Interior. 1998. *Status and Trends of the Nation's Biological Resources.* Reston, VA: USGS.

U.S. Geological Survey, U.S. Department of the Interior. 2001. "USGS Utah Water Science Center, Great Salt Lake." http://ut.water.usgs.gov.

U.S. Geological Survey, U.S. Department of the Interior. 2006. "Species Checklist." Jamestown, ND: Northern Prairie Wildlife Research Center. Checklist prepared for the prairie pothole region of Minnesota. Northern Prairie Wildlife Research Center Online. http://www.npwrc.usgs.gov.

Wiebe, K., and N. Gollehon, eds. 2006. "Agricultural Resources and Environmental Indicators." Washington, DC: U.S. Department of Agriculture, Economic Research Service. http://www.ers.usda.gov/publications/arei/eib16.

Vannote, R. L., G. W. Minshall, K. W. Cummings, J. R. Sedell, and C. E. Cushing. 1980. "The River Continuum Concept." *Canadian Journal of Fisheries and Aquatic Sciences* 37 (1): 130–137.

Vartapedian, B. B., and M. B. Jackson. 1997. "Plant Adaptations to Anaerobic Stress." *Annals of Botany* 79 (Supp. A): 3–20.

Vyruchalkina, T. Y. 2004. "Lake Baikal and the Angara River Before and After the Construction of Reservoirs." *Water Resources* 31 (5): 483–489. Translated from *Vodnye Resursy* 31 (5): 526–532.

Weller, M. W. 1999. *Wetland Birds. Habitat Resources and Conservation Implications.* New York: Cambridge University Press.

Welsch, D. J., D. L. Smart, J. N. Boyer, P. Minkin, H. C. Smith, and T. L. McCandless. 1995. "Forested wetlands. Functions, benefits and the Use of Best Management Practices." NA-PR-01-95. Radnor, PA: USDA Forest Service. http://www.na.fs.fed.us/spfo/pubs/n_resource/wetlands.

Welsh, S. A., D. A. Cincotta, and J. F. Switzer. 2006. "Fishes of the Bluestone National Scenic River." Natural Resources Technical Report NPS/NER/NRTR-2006/049. Philadelphia: National Park Service Regional Office, U.S. Department of the Interior.

Wetzel, R. G. 1983. *Limnology*, 2nd ed. New York: Saunders College Press.

Williams, C. E., R. D. Bivens, and B. D. Carter. 2006. "A Status Survey of the Obey Crayfish (*Cambarus obeyensis*)." Nashville: Tennessee Wildlife Resources Agency.

Windsor, B. 2000. *Tennessee River Basin Freshwater Mussels.* Gloucester, VA: U.S. Fish and Wildlife Service.

World Wildlife Fund. 2001. *Terrestrial Ecoregions of North America: A Conservation Assessment.* Washington, DC: Island Press, 2001. Ecoregion descriptions. http://www.worldwildlife.org/science/ecoregions.cfm.

Yount, J. D., and J. G. Niemi. 1990. "Recovery of Lotic Communities and Ecosystems from Disturbance—a Narrative Review of Case Studies." *Environmental Management* 14 (5): 547–569.

Index

Abyssal zone, lakes, 151, 162, 164

Acid-base conditions. *See* pH conditions

Adaptations: anaerobic conditions, 89, 151; anseriformes, 21; charadriiformes, 22; chemical properties of water, 7–13, 14; conditions requiring, general, 4–13; evolution and, xii, 36; fish, 19; general, vii, 5; physical properties of water, general, 4–7; river flows, 26, 37–38; salinity, 8–10, 97–98, 153–56; salt lakes, 153–56; water temperature, 163; water velocity, 14, 16, 26, 37–38. *See also* Disturbance; Human impacts

Alderflies, dobsonflies, and fishflies (Megaloptera), 18, 42

Algae: cryophilic, 63; floodplain lakes, 63; general, 13–14; Lake Baikal, 163; Lake Ontario, 187; lakes, classification of, 143; Lake Victoria, 171; Pantanal, 116; prairie potholes, 112; rivers, 39–40, 53, 64; water density, 5; water temperature, 7. *See also* Periphyton; Trophic status, lakes

Alien species. *See* Invasive/introduced species

Amazon River: biota, 53–60; floodplain, 52, 53, 54–55; floods, 52–56 passim; geology, 50, 51–52; headwaters, 49, 51–52, 55; human impacts, 59–60; overview, 27, 49–60, 80–82, 107; phytoplankton, 53, 54; problems and prospects, 59–60; rainforest, 49–55 passim; sediments, 49–50, 51, 53, 54–55; tributaries, 51, 52, 55–59 passim, 89; wetland tributaries, 100, 107

Amphibians: general, 20; lakes, 172, 173; Pantanal, 117, 120; playa lakes, 113; riverine, 41, 59, 73; wetlands, 95–96, 105–6

Amphipods: general, 16; Lake Ontario, 189, 190, 196; lakes, other, 163, 164, 204, 207, 208; rivers, 38, 42; wetlands, 100, 118

Amur River, 60–65, 82–83

Anaerobic conditions: adaptations to, 89, 151; eutrophic lakes, 11, 149, 159, 168–80; hydric soils, wetlands, 86, 89–90, 94, 211; manmade lakes, 159; peatlands, 108–10, 133; wetland bacteria, 93

Annelid worms, 15, 41, 207

Anthropogenic impacts. *See* Human impacts

About the Author

RICHARD A. ROTH received a B.A. in economics from the University of Virginia in 1972, and a master's degree in urban and regional planning (1989) and a Ph.D. in environmental design and planning from Virginia Polytechnic Institute and State University (1993). His academic interests include sustainable communities, environmental planning, and water resources. An avid whitewater paddler and self-described "river rat," Dr. Roth lives in Blacksburg, Virginia, with his wife, Polly Jones, and daughter Liza.